AN INTRODUCTION TO PARTICLE PHYSICS

Edited by
Deah E. Elmundo

Delve Publishing

An Introduction to Particle Physics

Edited by Deah E. Elmundo

Edition 2018

Delve Publishing

Published by

Delve Publishing LLC

708 3rd Avenue, 6th Floor

New York

NY 10801

United States of America

www.delvepublishing.com

ISBN: 978-1-68095-807-2 (Hardback)

Printed in United States of America

Library of Congress Control Number: 2016920690

About the Editor

Deah E. Elmundo

Deah obtained her BS in Physics for Teachers from the Philippine Normal University where she graduated Magna cum laude and completed her academic requirements leading to MS in Teaching Physics from De La Salle University. She loves teaching Classical and Modern Physics and is presently a full-time science faculty of a junior high school in her country. She is also a maths and physics content specialist to a few tuition companies in Singapore and Australia.

Table of Contents

3 FERMIONS

4 BOSONS

5 HADRONS

6 FADDEEV–POPOV GHOSTS

7 BARYONS AND HYPERONS

8 PHENOMENA OF QUASIPARTICLE

List of Abbreviations

ALC	Avoided Level Crossing
ESR	Electron Spin Resonance
ILC	International Linear Collider
NMR	Nuclear Magnetic Resonance
NRQED	Non-Relativistic Quantum Electrodynamics
QCD	Quantum Chromodynamics
QED	Quantum Electrodynamics
SLAC	Stanford Linear Accelerator Center
VMD	Vector Meson Dominance

Preface

Protons, electrons, neutrons, neutrinos and even quarks are often featured in news of scientific discoveries. All of these, and a whole "zoo" of others, are tiny sub-atomic particles too small to be seen even in microscopes. While molecules and atoms are the basic elements of familiar substances that we can see and feel, we have to "look" within atoms in order to learn about the "elementary" sub-atomic particles and to understand the nature of our Universe. The science of this study is called Particle Physics, Elementary Particle Physics or sometimes High Energy Physics (HEP). Atoms were postulated long ago by the Greek philosopher Democritus, and until the beginning of the 20th century, atoms were thought to be the fundamental indivisible building blocks of all forms of matter. Protons, neutrons and electrons came to be regarded as the fundamental particles of nature when we learned in the 1900's through the experiments of Rutherford and others that atoms consist of mostly empty space with electrons surrounding a dense central nucleus made up of protons and neutrons.

Modern particle physics research is focused on subatomic particles, including atomic constituents such as electrons, protons, and neutrons (protons and neutrons are composite particles called baryons, made of quarks), produced by radioactive and scattering processes, such as photons, neutrinos, and muons, as well as a wide range of exotic particles. Dynamics of particles is also governed by quantum mechanics; they exhibit wave–particle duality, displaying particle-like behaviour under certain experimental conditions and wave-like behaviour in others. In more technical terms, they are described by quantum state vectors in a Hilbert space, which is also treated in quantum field theory. Following the convention of particle physicists, the term elementary particles is applied to those particles that are, according to current understanding, presumed to be indivisible and not composed of other particles. All particles and their interactions observed to date can be described almost entirely by a quantum field theory called the Standard Model.

Content Coverage

Chapters One and *Two* present an overview of particle physics and atoms with atomic nucleus, correspondingly. Atoms are the basic units of matter and the defining structure of elements; atoms are made up of three particles: protons, neutrons and electrons.

Chapters Three and *Four* present a coverage of fermions satellite and GPS, consistently. In particle physics, a fermion is any particle characterized by Fermi–Dirac statistics, whereas a boson is a particle that follows Bose–Einstein statistics. Bosons make up one of the two classes of particles, the other being fermions.

Chapters Five and *Six* are intended to focus on properties of hadrons and Faddeev–Popov ghosts, separately. In particle physics, a hadron is a composite particle made of quarks held together by the strong force in a similar way as molecules are held together by the electromagnetic force. However, the necessity for Faddeev–Popov ghosts follows from the requirement that quantum field theories yield unambiguous, non-singular solutions.

Chapters Seven and *Eight* explain the fundamentals of baryons and quasiparticles, separately. Baryons are strongly interacting fermions, that is, they are acted on by the strong nuclear force and are described by Fermi–Dirac Statistics, which apply to all particles obeying the Pauli Exclusion Principle. The quasiparticle concept is most important in condensed matter physics since it is one of the few known ways of simplifying the quantum mechanical many-body problem.

CHAPTER 1

Particle Physics: An Overview

Chapter Contents

- **Introduction**
- **Subatomic Particles**
- **History**
- **Experimental Laboratories**
- **Theory**

Particle physics is a branch of physics that studies the elementary constituents of matter and radiation, and the interactions between them.

It is also called "high energy physics", because many elementary particles do not occur under normal circumstances in nature, but can be created and detected during energetic collisions of other particles, as is done in particle accelerators.

INTRODUCTION

Particle physics (also high energy physics) is the branch of physics that studies the nature of the particles that constitute *matter* (particles with mass) and *radiation* (massless particles). Although the word "particle" can refer to various types of very small objects (e.g. protons, gas particles, or even household dust), "particle physics" usually investigates the irreducibly smallest detectable particles and the irreducibly fundamental force fields necessary to explain them. By our current understanding, these elementary particles are excitations of the quantum fields that also govern their interactions. The currently dominant theory explaining these fundamental particles and fields, along with their dynamics, is called the Standard Model. Thus, modern particle physics generally investigates the Standard Model and its various possible extensions, e.g. to the newest "known" particle, the Higgs boson, or even to the oldest known force field, gravity.

SUBATOMIC PARTICLES

Modern particle physics research is focused on subatomic particles, including atomic constituents such as electrons, protons, and neutrons (protons and neutrons are composite particles called baryons, made of quarks), produced by radioactive and scattering processes, such as photons, neutrinos, and muons, as well as a wide range of exotic particles. Dynamics of particles is also governed by quantum mechanics; they exhibit wave–particle duality, displaying particle-like behavior under certain experimental conditions and wave-like behavior in others. In more technical terms, they are described by quantum state vectors in a Hilbert space, which is also treated in quantum field theory. Following the convention of particle physicists, the term *elementary particles* is applied to those particles that are, according to current understanding, presumed to be indivisible and not composed of other particles.

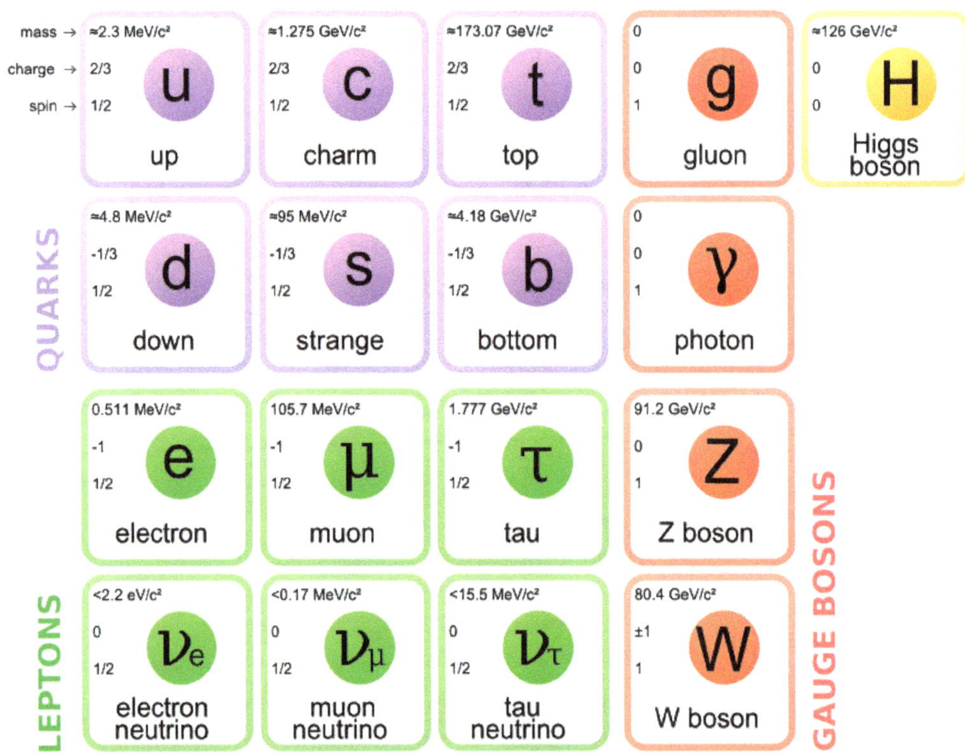

The particle content of the Standard Model of Physics.

Elementary Particles					
	Types	**Generations**	**Antiparticle**	**Colors**	**Total**
Quarks	2	3	Pair	3	36
Leptons			Pair	None	12
Gluons	1	1	Own	8	8
Photon			Own	None	1
Z Boson			Own		1
W Boson			Pair		2
Higgs			Own		1
Total number of (known) elementary particles:					**61**

All particles and their interactions observed to date can be described almost entirely by a quantum field theory called the Standard Model. The Standard Model, as currently formulated, has 61 elementary particles. Those elementary particles can combine to form composite particles, accounting for the hundreds of other species of particles that have been discovered since the 1960s. The Standard Model has been found to agree with almost all the experimental tests conducted to date. However, most particle physicists believe that it is an incomplete description of nature and that a more fundamental theory awaits discovery. In recent years, measurements of neutrino mass have provided the first experimental deviations from the Standard Model.

HISTORY

The idea that all matter is composed of elementary particles dates from at least the 6th century BC. In the 19th century, John Dalton, through his work on stoichiometry, concluded that each element of nature was composed of a single, unique type of particle. The word *atom*, after the Greek word *atomos* meaning "indivisible", has since then denoted the smallest particle of a chemical element, but physicists soon discovered that atoms are not, in fact, the fundamental particles of nature, but are conglomerates of even smaller particles, such as the electron. The early 20th century explorations of nuclear physics and quantum physics led to proofs of nuclear fission in 1939 by Lise Meitner (based on experiments by Otto Hahn), and nuclear fusion by Hans Bethe in that same year; both discoveries also led to the development of nuclear weapons. Throughout the 1950s and 1960s, a bewildering variety of particles were found in scattering experiments. It was referred to informally as the "particle zoo". That term was deprecated after the formulation of the Standard Model during the 1970s, in which the large number of particles was explained as combinations of a (relatively) small number of more fundamental particles.

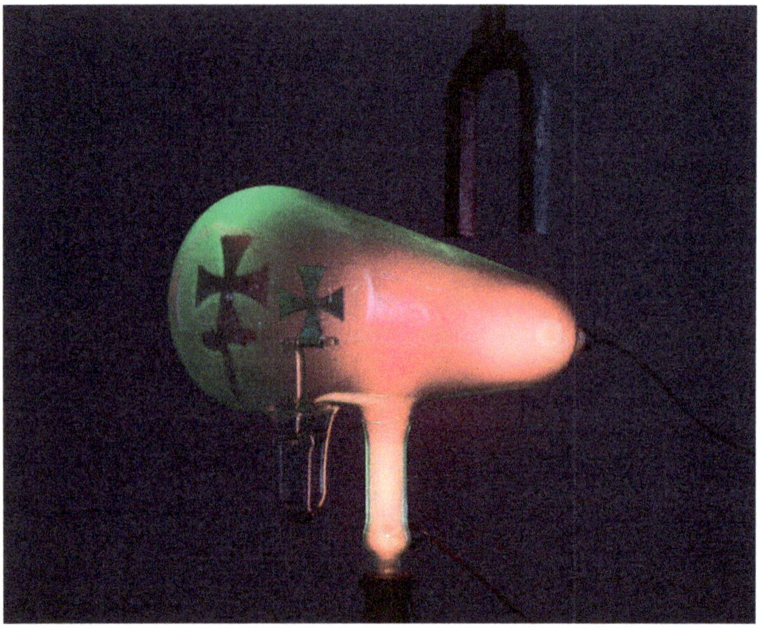

A Crookes tube with a magnetic deflector.

Early Development

The idea that all matter is composed of elementary particles dates to at least the 6th century BC. The philosophical doctrine of atomism and the nature of elementary particles were studied by ancient Greek philosophers such as Leucippus, Democritus, and Epicurus; ancient Indian philosophers such as Kanada, Dignāga, and Dharmakirti; Muslim scientists such as Ibn al-Haytham, Ibn Sina, and Mohammad al-Ghazali; and in early modern Europe by physicists such as Pierre Gassendi, Robert Boyle, and Isaac Newton. The particle theory of light was also proposed by Ibn al-Haytham, Ibn Sina, Gassendi, and Newton.

Those early ideas were founded through abstract, philosophical reasoning rather than experimentation and empirical observation and represented only one line of thought among many. In contrast, certain ideas of Gottfried Wilhelm Leibniz contradict to almost everything known in modern physics.

From Atoms to Nucleons

First Subatomic Particles

However, near the end of 19th century, physicists discovered that Dalton's atoms are not, in fact, the fundamental particles of nature, but conglomerates of even smaller particles. Electron was discovered between 1879 and 1897 in works of William Crookes, Arthur Schuster, J. J. Thomson, and other physicists; its charge was carefully measured by Robert Andrews Millikan and Harvey Fletcher in their oil drop experiment of 1909. Physicists theorized that negatively charged electrons are constituent part of "atoms", along with some (yet unknown) positively charged substance, and it was later confirmed. Electron became the first elementary, truly fundamental particle discovered.

Studies of the "radioactivity", that soon revealed the phenomenon of radioactive decay, provided another argument against considering chemical elements as fundamental nature's elements. Despite these discoveries, the term *atom* stuck to Dalton's (chemical) atoms and now denotes the smallest particle of a chemical element, not something really indivisible.

Researching Particles' Interaction

Early 20th-century physicists knew only two fundamental forces: electromagnetism and gravitation, where the latter could not explain the structure of atoms. So, it was obvious to assume that unknown positively charged substance attracts electrons by Coulomb force.

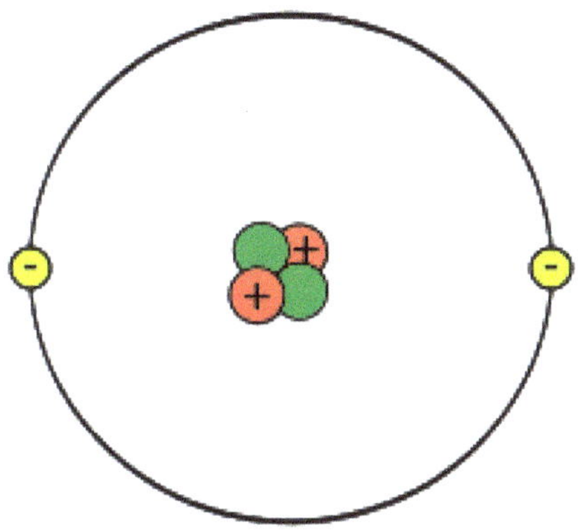

In 1909 Ernest Rutherford and Thomas Royds demonstrated that an alpha particle combines with two electrons and forms a helium atom. In modern terms, alpha particles are doubly ionized helium (more precisely, ^{4}He) atoms. Speculation about the structure of atoms was severely constrained by Rutherford's 1907 gold foil experiment, showing that the atom is mainly empty space, with almost all its mass concentrated in a tiny atomic nucleus.

Inside the Atom

By 1914, experiments by Ernest Rutherford, Henry Moseley, James Franck and Gustav Hertz had largely established the structure of an atom as a dense nucleus of positive charge surrounded by lower-mass electrons. These discoveries shed a light to the nature of radioactive decay and other forms of transmutation of elements, as well as of elements themselves. It appeared that atomic number is nothing else than (positive) electric charge of the atomic nucleus of a particular atom. Chemical transformations, governed by electromagnetic interactions, do not change nuclei – that's why elements are chemically indestructible. But when the nucleus change its charge and/or mass (by emitting or capturing a particle), the atom can become the one of another element. Special relativity explained how the *mass defect* is related to the energy produced or consumed in reactions. The branch of physics that studies transformations and the structure of nuclei is now called nuclear physics, contrasted to atomic physics that studies the structure and properties of atoms ignoring most nuclear aspects. The development in the nascent quantum physics, such as Bohr model, led to the understanding of chemistry in terms of the arrangement of electrons in the mostly empty volume of atoms.

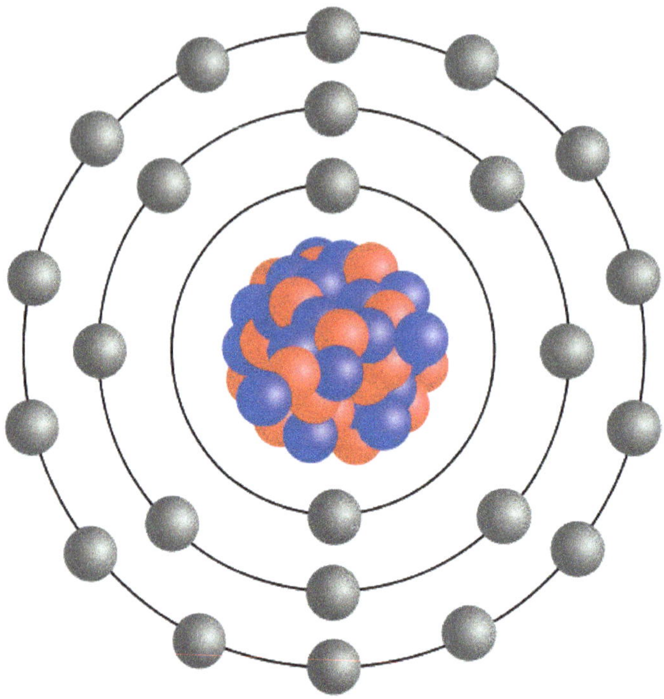

In 1918, Rutherford confirmed that the hydrogen nucleus was a particle with a positive charge, which he named the proton. By then, Frederick Soddy's researches of radioactive elements, and experiments of J. J. Thomson and F.W. Aston conclusively demonstrated existence of isotopes, whose nuclei have different masses in spite of identical atomic numbers. It prompted Rutherford to conjecture that all nuclei other than hydrogen contain chargeless particles, which he named the neutron. Evidences that atomic nuclei consist of some smaller particles (now called *nucleons*) grew; it became obvious that, while protons repulse each other electrostatically, nucleons attract each other by some new force (nuclear force). It culminated in proofs of nuclear fission in 1939 by Lise Meitner (based on experiments by Otto Hahn), and nuclear fusion by Hans Bethe in that same year. Those discoveries gave rise to an active industry of generating one atom from another, even rendering possible (although it will probably never be profitable) the transmutation of lead into gold; and, those same discoveries also led to the development of nuclear weapons.

Revelations of Quantum Mechanics

Further understanding of atomic and nuclear structures became impossible without improving the knowledge about the essence of particles. Experiments and improved theories (such as Erwin Schrödinger's "electron waves") gradually revealed that there is no fundamental difference between particles and waves. For example, electromagnetic waves were reformulated in terms of particles called *photons*. It also revealed that physical objects do not change their parameters, such as total energy, position and momentum, as continuous functions of time, as it was thought of in classical physics.

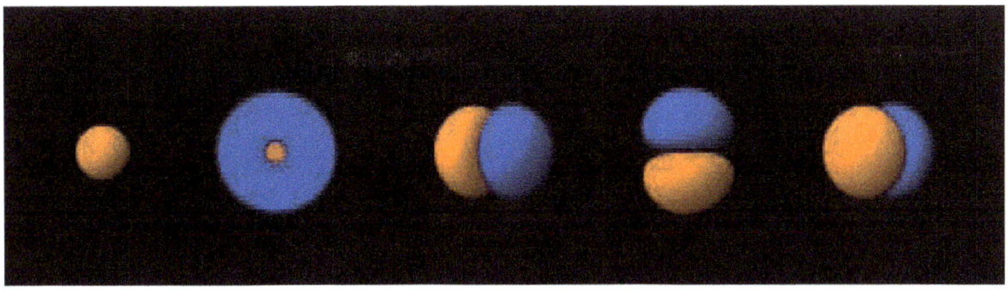

Atomic orbitals of Period 2 elements:1s 2s 2p (3 items). All complete subshells (including 2p) are inherently spherically symmetric, but it is convenient to assign to "distinct" p-electrons these two-lobed shapes.

Another crucial discovery was identical particles or, more generally, quantum particle statistics. It was established that all electrons are identical: although two or more electrons can exist simultaneously that have different parameters, but they do not keep separate, distinguishable histories. This also applies to protons, neutrons, and (with certain differences) to photons as well. It suggested that there is a limited number of sorts of smallest particles in the universe.

The spin–statistics theorem established that any particle in our spacetime may be either a boson (that means its statistics is Bose–Einstein) or a fermion (that means its statistics is Fermi–Dirac). It was later found that all fundamental bosons transmit forces, like the photon that transmits light. Some of non-fundamental bosons (namely, mesons) also may transmit forces (see below), although non-fundamental ones. Fermions are particles "like electrons and nucleons" and generally comprise the matter. Note that any subatomic or atomic particle composed of even *total* number of fermions (such as protons, neutrons, and electrons) is a boson, so a boson is not necessarily a force transmitter and perfectly can be an ordinary material particle.

The spin is the quantity that distinguishes bosons and fermions. Practically it appears as an intrinsic angular momentum of a particle, that is unrelated to its motion but is linked with some other features like a magnetic dipole. Theoretically it is explained from different types representations of symmetry groups, namely tensor representations (including vectors and scalars) for bosons with their integer (in $\hbar$) spins, and spinor representations for fermions with their half-integer spins.

This culminated in the formulation of ideas of a quantum field theory. The first (and the only mathematically complete) of these theories, quantum electrodynamics, allowed to explain thoroughly the structure of atoms, including the Periodic Table and atomic spectra. Ideas of quantum mechanics and quantum field theory were applied to nuclear physics too. For example, α decay was explained as a quantum tunneling through nuclear potential, nucleons' fermionic statistics explained the nucleon pairing, and Hideki

Yukawa proposed certain virtual particles (now knows as π-mesons) as an explanation of the nuclear force.

Inventory

Modern Nuclear Physics

Development of nuclear models (such as the liquid-drop model and nuclear shell model) made prediction of properties of nuclides possible. No existing model of nucleon–nucleon interaction can *analytically* compute something more complex than ^{4}He based on principles of quantum mechanics, though (note that complete computation of electron shells in atoms is also impossible yet).

The most developed branch of nuclear physics in 1940s was studies related to nuclear fission due to its military significance. The main focus of fission-related problems is interaction of atomic nuclei with neutrons: a process that occurs in a fission bomb and a nuclear fission reactor. It gradually drifted away from the rest of subatomic physics and virtually became the nuclear engineering. First synthesized transuranium elements were also obtained in this context, through neutron capture and subsequent β^{-} decay.

The elements beyond fermium cannot be produced in this way. To make a nuclide with more than 100 protons per nucleus one has to use an inventory and methods of particle physics, namely to accelerate and collide atomic nuclei. Production of progressively heavier synthetic elements continued into 21st century as a branch of nuclear physics, but only for scientific purposes.

The third important stream in nuclear physics are researches related to nuclear fusion. This is related to thermonuclear weapons (and conceived peaceful thermonuclear energy), as well as to astrophysical researches, such as stellar nucleosynthesis and Big Bang nucleosynthesis.

Strange Particles and Mysteries of the Weak Interaction

In the 1950s, with development of particle accelerators and studies of cosmic rays, inelastic scattering experiments on protons (and other atomic nuclei) with energies about hundreds of MeVs became affordable. They created some short-lived resonance "particles", but also *hyperons* and *K-mesons* with unusually long lifetime. The cause of the latter was found in a new quasi-conserved quantity, named *strangeness*, that is conserved in all circumstances except for the weak interaction. The strangeness of heavy particles and the μ-lepton were first two signs of what is now known as the second generation of fundamental particles.

The weak interaction revealed soon yet another mystery. In 1957 it was found that it does not conserve parity. In other words, the mirror symmetry was disproved as a fundamental symmetry law.

Throughout the 1950s and 1960s, improvements in particle accelerators and particle detectors led to a bewildering variety of particles found in high-energy experiments. The term *elementary particle* came to refer to dozens of particles, most of them unstable. It prompted Wolfgang Pauli's remark: "Had I foreseen this, I would have gone into botany". The entire collection was nicknamed the "particle zoo". It became evident that some smaller constituents, yet invisible, form mesons and baryons that counted most of then-known particles.

Deeper Constituents of Matter

The interaction of these particles by scattering and decay provided a key to new fundamental quantum theories. Murray Gell-Mann and Yuval Ne'eman brought some order to mesons and baryons, the most numerous classes of particles, by classifying them according to certain qualities. It began with what Gell-Mann referred to as the "Eightfold Way", but proceeding into several different "octets" and "decuplets" which could predict new particles, most famously the Ω−, which was detected at Brookhaven National Laboratory in 1964, and which gave rise to the *quark* model of hadron composition. While the quark model at first seemed inadequate to describe strong nuclear forces, allowing the temporary rise of competing theories such as the S-matrix theory, the establishment of quantum chromodynamics in the 1970s finalized a set of fundamental and exchange particles (Kragh 1999). It postulated the fundamental strong interaction, experienced by quarks and mediated by gluons. These particles were proposed as a building material for hadrons. This theory is unusual because individual (free) quarks cannot be observed, unlike the situation with composite atoms where electrons and nuclei can be isolated by transferring ionization energy to the atom.

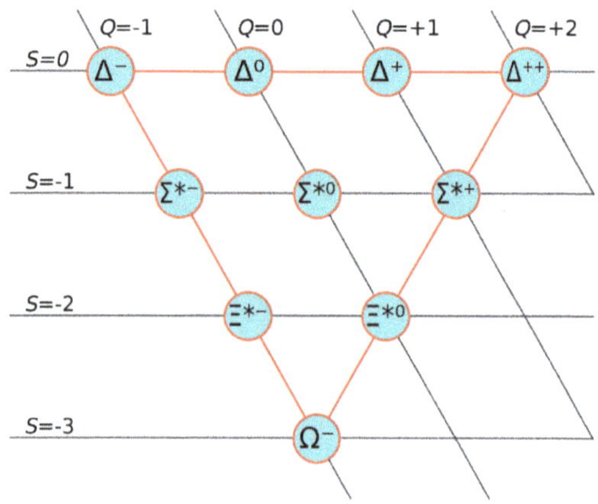

Classification of spin-3/2 baryons known in 1960s

Then, the old, broad denotation of the term *elementary particle* was deprecated and a replacement term *subatomic particle* covered all the "zoo", with its hyponym "hadron" referring to composite particles directly explained by the quark model. The designation of an "elementary" (or "fundamental") particle was reserved for leptons, quarks, their antiparticles, and quanta of fundamental interactions only.

Quarks, Leptons, and Four Fundamental Forces

Because the quantum field theory postulates no difference between particles and interactions, classification of elementary particles allowed also to classify interactions and fields.

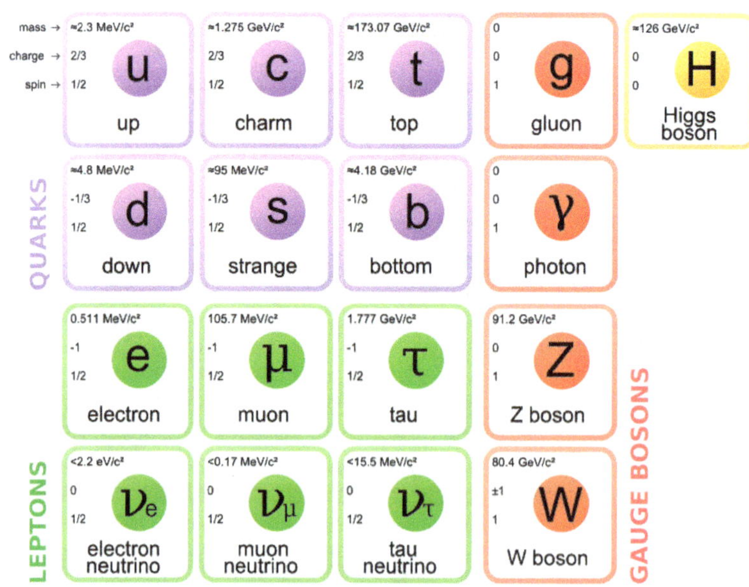

The Standard Model

Now a large number of particles and (non-fundamental) interactions is explained as combinations of a (relatively) small number of fundamental substances, thought to be fundamental interactions (incarnated in fundamental bosons), quarks (including antiparticles), and leptons (including antiparticles). As the theory distinguished *several* fundamental interactions, it became possible to see which elementary particles participate in which interaction. Namely:

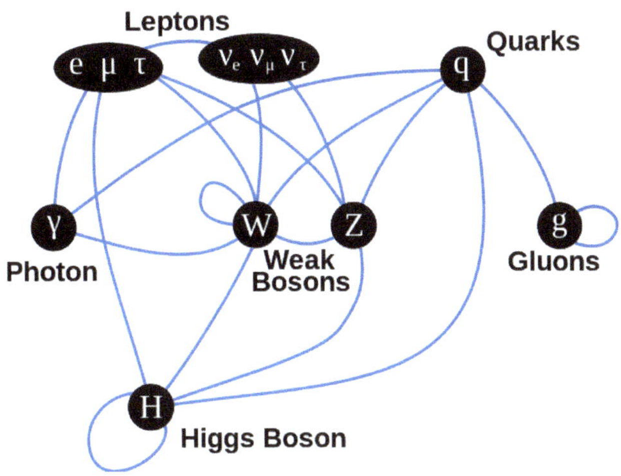

- All particles participate in gravitation.
- All charged elementary particles participate in electromagnetic interaction.
- As a consequence, neutron participate in it with its magnetic dipole in spite of zero electric charge. This is because it is composed of *charged* quarks whose charges sum to zero.
- All fermions participate in the weak interaction.
- Quarks participate in the strong interaction, along gluons (its own quanta), but not leptons nor any fundamental bosons other than gluons.

The next step was a reduction in number of fundamental interactions, envisaged by early 20th century physicists as the "united field theory". The first successful modern unified theory was the electroweak theory, developed by Abdus Salam, Steven Weinberg and, subsequently, Sheldon Glashow. This development culminated in the completion of the theory called the Standard Model in the 1970s, that included also the strong interaction, thus covering three fundamental forces. After the discovery, made at CERN, of the existence of neutral weak currents, mediated by the Z boson foreseen in the standard model, the physicists Salam, Glashow and Weinberg received the 1979 Nobel Prize in Physics for their electroweak theory. The discovery of the weak gauge bosons (quanta of the weak interaction) through the 1980s, and the verification of their properties through the 1990s is considered to be an age of consolidation in particle physics.

While accelerators have confirmed most aspects of the Standard Model by detecting expected particle interactions at various collision energies, no theory reconciling general relativity with the Standard Model has yet been found, although supersymmetry and string theory were believed by many theorists to be a promising avenue forward. The

Large Hadron Collider, however, which began operating in 2008, has failed to find any evidence whatsoever that is supportive of supersymmetry and string theory, and appears unlikely to do so, meaning “the current situation in fundamental theory is one of a serious lack of any new ideas at all.” This state of affairs should not be viewed as a crisis in physics, but rather, as David Gross has said, “the kind of acceptable scientific confusion that discovery eventually transcends.”

The fourth fundamental force, gravitation, is not yet integrated into particle physics in a consistent way.

Higgs Boson

As of 2011, the Higgs boson, the quantum of a field that is thought to provide particles with rest masses, remained the only particle of the Standard Model to be verified. On July 4, 2012, physicists working at CERN’s Large Hadron Collider announced that they had discovered a new subatomic particle greatly resembling the Higgs boson, a potential key to an understanding of why elementary particles have masses and indeed to the existence of diversity and life in the universe. Rolf-Dieter Heuer, the director general of CERN from 2009 to 2015, said that it was too soon to know for sure whether it is an entirely new particle, which weighs in at 125 billion electron volts – one of the heaviest subatomic particles yet – or, indeed, the elusive particle predicted by the Standard Model, the theory that has ruled physics for the last half-century. It is unknown if this particle is an impostor, a single particle or even the first of many particles yet to be discovered.

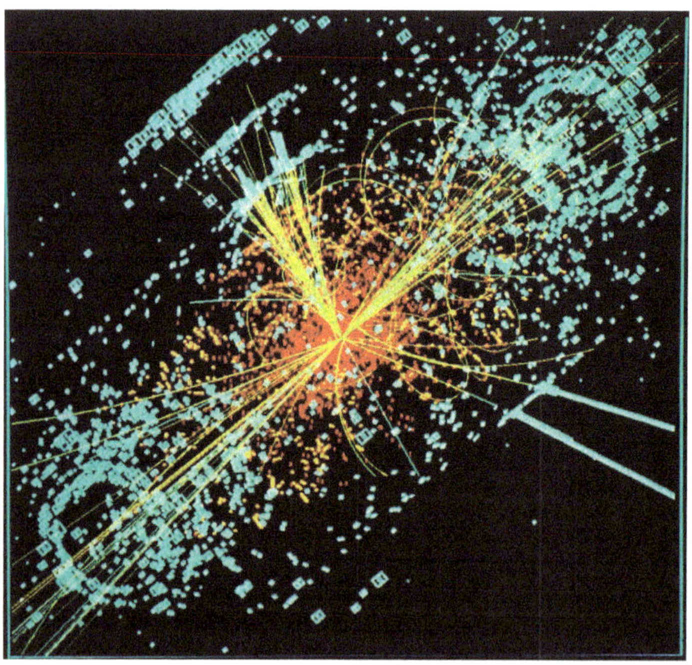

One possible signature of a Higgs boson from a simulated proton–proton collision. It decays almost immediately into two jets of hadrons and two electrons, visible as lines.

The latter possibilities are particularly exciting to physicists since they could point the way to new deeper ideas, beyond the Standard Model, about the nature of reality. For now, some physicists are calling it a "Higgs–like" particle. Joe Incandela, of the University of California, Santa Barbara, said, "It's something that may, in the end, be one of the biggest observations of any new phenomena in our field in the last 30 or 40 years, going way back to the discovery of quarks, for example." The groups operating the large detectors in the collider said that the likelihood that their signal was a result of a chance fluctuation was less than one chance in 3.5 million, so-called "five sigma," which is the gold standard in physics for a discovery. Michael Turner, a cosmologist at the University of Chicago and the chairman of the physics center board, said

This is a big moment for particle physics and a crossroads — will this be the high water mark or will it be the first of many discoveries that point us toward solving the really big questions that we have posed?

—Michael Turner, University of Chicago

Confirmation of the Higgs boson or something very much like it would constitute a rendezvous with destiny for a generation of physicists who have believed the boson existed for half a century without ever seeing it. Further, it affirms a grand view of a universe ruled by simple and elegant and symmetrical laws, but in which everything interesting in it being a result of flaws or breaks in that symmetry. According to the Standard Model, the Higgs boson is the only visible and particular manifestation of an invisible force field that permeates space and imbues elementary particles that would otherwise be massless with mass. Without this Higgs field, or something like it, physicists say all the elementary forms of matter would zoom around at the speed of light; there would be neither atoms nor life. The Higgs boson achieved a notoriety rare for abstract physics. To the eternal dismay of his colleagues, Leon Lederman, the former director of Fermilab, called it the "God particle" in his book of the same name, later quipping that he had wanted to call it "the goddamn particle". Professor Incandela also stated,

This boson is a very profound thing we have found. We're reaching into the fabric of the universe at a level we've never done before. We've kind of completed one particle's story [...] We're on the frontier now, on the edge of a new exploration. This could be the only part of the story that's left, or we could open a whole new realm of discovery.

—Joe Incandela, University of California

In quantum theory, which is the language of particle physicists, elementary particles are divided into two rough categories: fermions, which are bits of matter like electrons, and bosons, which are bits of energy and can transmit forces, like the photon that transmits light. Dr. Peter Higgs was one of six physicists, working in three independent groups, who in 1964 invented the notion of the cosmic molasses, or Higgs field. The others were Tom Kibble of Imperial College, London; Carl Hagen of the University of Rochester; Gerald Guralnik of Brown University; and François Englert and Robert Brout, both of Université Libre de Bruxelles. One implication of their theory was that this Higgs field, normally invisible and, of course, odorless, would produce its own quantum particle if hit hard enough, by the right amount of energy. The particle would be fragile and fall apart within a millionth of a second in a dozen different ways depending upon its own mass. Unfortunately, the theory did not say how much this particle should weigh, which is what made it so difficult to find. The particle eluded researchers at a succession of particle accelerators, including the Large Electron–Positron Collider at CERN, which closed down in 2000, and the Tevatron at the Fermi National Accelerator Laboratory, or Fermilab, in Batavia, Ill., which shut down in 2011.

Further experiments continued and in March 2013 it was tentatively confirmed that the newly discovered particle was a Higgs Boson.

Although they have never been seen, Higgslike fields play an important role in theories of the universe and in string theory. Under certain conditions, according to the strange accounting of Einsteinian physics, they can become suffused with energy that exerts an antigravitational force. Such fields have been proposed as the source of an enormous burst of expansion, known as inflation, early in the universe and, possibly, as the secret of the dark energy that now seems to be speeding up the expansion of the universe.

Further Theoretical Development

Modern theoretical development includes refining of the Standard Model, researching in its foundations such as the Yang–Mills theory, and researches in computational methods such as the lattice QCD.

A long-standing problem is quantum gravitation. No solution that is useful for particle physics has been achieved.

Further Experimental Development

There are researches about quark–gluon plasma, a new (hypothetical) state of matter. There are also some recent experimental evidences that tetraquarks and glueballs exist.

The proton decay is not observed (or, generally, non-conservation of the baryon number), but predicted by the Standard Model, so there are searches for it.

Standard Model

The current state of the classification of all elementary particles is explained by the Standard Model. It describes the strong, weak, and electromagnetic fundamental interactions, using mediating gauge bosons. The species of gauge bosons are the gluons, W^-, W^+ and Z bosons, and the photons. The Standard Model also contains 24 fundamental particles (12 particles and their associated anti-particles), which are the constituents of all matter. Finally, the Standard Model also predicted the existence of a type of boson known as the Higgs boson. Early in the morning on 4 July 2012, physicists with the Large Hadron Collider at CERN announced they had found a new particle that behaves similarly to what is expected from the Higgs boson.

EXPERIMENTAL LABORATORIES

The world's major particle physics laboratories are:

- Brookhaven National Laboratory (Long Island, United States). Its main facility is the Relativistic Heavy Ion Collider (RHIC), which collides heavy ions such as gold ions and polarized protons. It is the world's first heavy ion collider, and the world's only polarized proton collider.

- Budker Institute of Nuclear Physics (Novosibirsk, Russia). Its main projects are now the electron-positron colliders VEPP-2000, operated since 2006, and VEPP-4, started experiments in 1994. Earlier facilities include the first electron-electron beam-beam collider VEP-1, which conducted experiments from 1964 to 1968; the electron-positron colliders VEPP-2, operated from 1965 to 1974; and, its successor VEPP-2M, performed experiments from 1974 to 2000.
- CERN (Organisation Europèene pour la Recherche Nucléaire) (Franco-Swiss border, near Geneva). Its main project is now the Large Hadron Collider (LHC), which had its first beam circulation on 10 September 2008, and is now the world's most energetic collider of protons. It also became the most energetic collider of heavy ions after it began colliding lead ions. Earlier facilities include the Large Electron–Positron Collider (LEP), which was stopped on 2 November 2000 and then dismantled to give way for LHC; and the Super Proton Synchrotron, which is being reused as a pre-accelerator for the LHC.
- DESY (Deutsches Elektronen-Synchrotron) (Hamburg, Germany). Its main facility is the Hadron Elektron Ring Anlage (HERA), which collides electrons and positrons with protons.
- Fermi National Accelerator Laboratory (Fermilab) (Batavia, United States). Its main facility until 2011 was the Tevatron, which collided protons and antiprotons and was the highest-energy particle collider on earth until the Large Hadron Collider surpassed it on 29 November 2009.
- Institute of High Energy Physics (IHEP) (Beijing, China). IHEP manages a number of China's major particle physics facilities, including the Beijing Electron Positron Collider (BEPC), the Beijing Spectrometer (BES), the Beijing Synchrotron Radiation Facility (BSRF), the International Cosmic-Ray Observatory at Yangbajing in Tibet, the Daya Bay Reactor Neutrino Experiment, the China Spallation Neutron Source, the Hard X-ray Modulation Telescope (HXMT), and the Accelerator-driven Sub-critical System (ADS) as well as the Jiangmen Underground Neutrino Observatory (JUNO).
- KEK (Tsukuba, Japan). It is the home of a number of experiments such as the K2K experiment, a neutrino oscillation experiment and Belle, an experiment measuring the CP violation of B mesons.
- SLAC National Accelerator Laboratory (Menlo Park, United States). Its 2-mile-long linear particle accelerator began operating in 1962 and was the basis for numerous electron and positron collision experiments until 2008. Since then the linear accelerator is being used for the Linac Coherent Light Source X-ray laser as well as advanced accelerator design research. SLAC staff continue to participate in developing and building many particle detectors around the world.

Many other particle accelerators also exist.

The techniques required for modern experimental particle physics are quite varied and complex, constituting a sub-specialty nearly completely distinct from the theoretical side of the field.

THEORY

Theoretical particle physics attempts to develop the models, theoretical framework, and mathematical tools to understand current experiments and make predictions for future experiments. There are several major interrelated efforts being made in theoretical particle physics today. One important branch attempts to better understand the Standard Model and its tests. By extracting the parameters of the Standard Model, from experiments with less uncertainty, this work probes the limits of the Standard Model and therefore expands our understanding of nature's building blocks. Those efforts are made challenging by the difficulty of calculating quantities in quantum chromodynamics. Some theorists working in this area refer to themselves as phenomenologists and they may use the tools of quantum field theory and effective field theory. Others make use of lattice field theory and call themselves *lattice theorists*.

Another major effort is in model building where model builders develop ideas for what physics may lie beyond the Standard Model (at higher energies or smaller distances). This work is often motivated by the hierarchy problem and is constrained by existing experimental data. It may involve work on supersymmetry, alternatives to the Higgs mechanism, extra spatial dimensions (such as the Randall-Sundrum models), Preon theory, combinations of these, or other ideas.

A third major effort in theoretical particle physics is string theory. *String theorists* attempt to construct a unified description of quantum mechanics and general relativity by building a theory based on small strings, and branes rather than particles. If the theory is successful, it may be considered a "Theory of Everything", or "TOE".

There are also other areas of work in theoretical particle physics ranging from particle cosmology to loop quantum gravity. This division of efforts in particle physics is reflected in the names of categories on the arXiv, a preprint archive: hep-th (theory), hep-ph (phenomenology), hep-ex (experiments), hep-lat (lattice gauge theory).

Practical Applications

In principle, all physics (and practical applications developed therefrom) can be derived from the study of fundamental particles. In practice, even if "particle physics" is taken to mean only "high-energy atom smashers", many technologies have been developed during these pioneering investigations that later find wide uses in society. Cyclotrons are used to produce medical isotopes for research and treatment (for example, isotopes used in PET imaging), or used directly for certain cancer treatments. The development of Superconductors has been pushed forward by their use in particle physics. The World Wide Web and touchscreen technology were initially developed at CERN.

Additional applications are found in medicine, national security, industry, computing, science, and workforce development, illustrating a long and growing list of beneficial practical applications with contributions from particle physics.

Future

The primary goal, which is pursued in several distinct ways, is to find and understand what physics may lie beyond the standard model. There are several powerful experimental reasons to expect new physics, including dark matter and neutrino mass. There are also theoretical hints that this new physics should be found at accessible energy scales.

Much of the effort to find this new physics are focused on new collider experiments. The Large Hadron Collider (LHC) was completed in 2008 to help continue the search for the Higgs boson, supersymmetric particles, and other new physics. An intermediate goal is the construction of the International Linear Collider (ILC), which will complement the LHC by allowing more precise measurements of the properties of newly found particles. In August 2004, a decision for the technology of the ILC was taken but the site has still to be agreed upon.

In addition, there are important non-collider experiments that also attempt to find and understand physics beyond the Standard Model. One important non-collider effort is the determination of the neutrino masses, since these masses may arise from neutrinos mixing with very heavy particles. In addition, cosmological observations provide many useful constraints on the dark matter, although it may be impossible to determine the exact nature of the dark matter without the colliders. Finally, lower bounds on the very long lifetime of the proton put constraints on Grand Unified Theories at energy scales much higher than collider experiments will be able to probe any time soon.

In May 2014, the Particle Physics Project Prioritization Panel released its report on particle physics funding priorities for the United States over the next decade. This report emphasized continued U.S. participation in the LHC and ILC, and expansion of the Long Baseline Neutrino Experiment, among other recommendations.

High Energy Physics Compared to Low Energy Physics

The term *high energy physics* requires elaboration. Intuitively, it might seem incorrect to associate "high energy" with the physics of very small, *low* mass objects, like subatomic particles. By comparison, an example of a macroscopic system, one gram of hydrogen, has $\sim 6\times10^{23}$ times the mass of a single proton. Even an entire beam of protons circulated in the LHC contains $\sim 3.23\times10^{14}$ protons, each with 6.5×10^{12} eV of energy, for a total beam energy of $\sim 2.1\times10^{27}$ eV or $\sim$ 336.4 MJ, which is still $\sim 2.7\times10^{5}$ times lower than the mass-energy of a single gram of hydrogen. Yet, the macroscopic realm is "low energy physics", while that of quantum particles is "high energy physics".

The interactions studied in other fields of physics and science have comparatively very low energy. For example, the photon energy of visible light is about 1.8 to 3.1 eV. Similarly, the bond-dissociation energy of a carbon-carbon bond is about 3.6 eV. Other chemical reactions typically involve similar amounts of energy. Even photons with far higher energy, gamma rays of the kind produced in radioactive decay, mostly have photon energy between 10^5 eV and 10^7 eV – still two orders of magnitude lower than the mass of a single proton. Radioactive decay gamma rays are considered as part of nuclear physics, rather than high energy physics.

The proton has a mass of around 9.4×10^{8} eV; some other massive quantum particles, both elementary and hadronic, have yet higher masses.

REFERENCES

1. http://home.web.cern.ch/topics/higgs-boson
2. http://www.nobelprize.org/nobel_prizes/physics/laureates/2013/advanced-physicsprize2013.pdf
3. Braibant, S.; Giacomelli, G.; Spurio, M. (2009). Particles and Fundamental Interactions: An Introduction to Particle Physics. Springer. pp. 313–314. ISBN 978-94-007-2463-1.
4. "Particle Physics and Astrophysics Research". The Henryk Niewodniczanski Institute of Nuclear Physics. Retrieved 31 May 2012.
5. "Fundamentals of Physics and Nuclear Physics" (PDF). Retrieved 21 July 2012.
6. "Scientific Explorer: Quasiparticles". Sciexplorer.blogspot.com. 22 May 2012. Retrieved 21 July 2012.
7. Nakamura, K (1 July 2010). "Review of Particle Physics". Journal of Physics G: Nuclear and Particle Physics. 37 (7A): 075021. Bibcode:2010JPhG...37g5021N. doi:10.1088/0954-3899/37/7A/075021.
8. Mann, Adam (28 March 2013). "Newly Discovered Particle Appears to Be Long-Awaited Higgs Boson - Wired Science". Wired.com. Retrieved 6 February 2014.
9. "Brookhaven National Laboratory – A Passion for Discovery". Bnl.gov. Retrieved 23 June 2012.
10. "index". Vepp2k.inp.nsk.su. Retrieved 21 July 2012.
11. "The VEPP-4 accelerating-storage complex". V4.inp.nsk.su. Retrieved 21 July 2012.
12. "VEPP-2M collider complex" (in Russian). Inp.nsk.su. Retrieved 21 July 2012.
13. "The Budker Institute Of Nuclear Physics". English Russia. 21 January 2012. Retrieved 23 June 2012.
14. "Welcome to". Info.cern.ch. Retrieved 23 June 2012.
15. "Germany's largest accelerator centre – Deutsches Elektronen-Synchrotron DESY". Desy.de. Retrieved 23 June 2012.
16. "Fermilab | Home". Fnal.gov. Retrieved 23 June 2012.
17. "IHEP | Home". ihep.ac.cn. Retrieved 29 November 2015.
18. "Kek | High Energy Accelerator Research Organization". Legacy.kek.jp. Retrieved 23 June 2012.
19. "SLAC National Accelerator Laboratory Home Page". Retrieved 19 February 2015.
20. arxiv.org
21. "Fermilab | Science at Fermilab | Benefits to Society". Fnal.gov. Retrieved 23

June 2012.

22. "CODATA Value: Avogadro constant". The NIST Reference on Constants, Units, and Uncertainty. US National Institute of Standards and Technology. June 2015. Retrieved 2016-12-10.
23. "Beam Requirements and Fundamental Choices" (PDF). CERN Engineering & Equipment Data Management Service (EDMS). Retrieved 10 December 2016.
24. "Fundamentals of Physics and Nuclear Physics" (PDF). Retrieved 2012-07-21.
25. "Scientific Explorer: Quasiparticles". Sciexplorer.blogspot.com. 2012-05-22. Retrieved 2012-07-21.
26. Smirnov, B.M. (2003). Physics of Atoms and Ions. Springer. pp. 14–21. ISBN 0-387-95550-X.
27. The discovery of the weak neutral currents, CERN courier, 2004-10-04, retrieved 2008-05-08
28. The Nobel Prize in Physics 1979, Nobel Foundation, retrieved 2008-09-10
29. Woit, Peter (20 October 2013). "Last Links For a While". Not Even Wrong. Retrieved 2 November 2013.
30. Peter Woit (28 May 2013). "A Tale of Two Oxford Talks". Not Even Wrong. Retrieved 19 October 2013.
31. Peter Byrne (24 May 2013). "Waiting for the Revolution". Quanta Magazine. simonsfoundation.org. Retrieved 19 October 2013.
32. http://www.nytimes.com/2012/07/05/science/cern-physicists-may-have-discovered-higgs-boson-particle.html?pagewanted=3&_r=1&ref=science
33. Rincon, Paul (2012-07-04). "BBC News - Higgs boson-like particle discovery claimed at LHC". Bbc.co.uk. Retrieved 2013-04-20.

CHAPTER 2

Atoms and Atomic Nucleus

Chapter Contents

Atoms are the basic units of matter and the defining structure of elements. Atoms are made up of three particles: protons, neutrons and electrons.

Protons and neutrons are heavier than electrons and reside in the center of the atom, which is called the nucleus. Electrons are extremely lightweight and exist in a cloud orbiting the nucleus. The electron cloud has a radius 10,000 times greater than the nucleus.

Protons and neutrons have approximately the same mass. However, one proton weighs more than 1,800 electrons. Atoms always have an equal number of protons and electrons, and the number of protons and neutrons is usually the same as well. Adding a proton to an atom makes a new element, while adding a neutron makes an isotope, or heavier version, of that atom.

INTRODUCTION

An atom is the smallest constituent unit of ordinary matter that has the properties of a chemical element. Every solid, liquid, gas, and plasma is composed of neutral or ionized atoms. Atoms are very small; typical sizes are around 100 picometers (a ten-billionth of a meter, in the short scale).

Atoms are small enough that attempting to predict their behavior using classical physics - as if they were billiard balls, for example - gives noticeably incorrect predictions due to quantum effects. Through the development of physics, atomic models have incorporated quantum principles to better explain and predict the behavior.

Every atom is composed of a nucleus and one or more electrons bound to the nucleus. The nucleus is made of one or more protons and typically a similar number of neutrons. Protons and neutrons are called nucleons. More than 99.94% of an atom's mass is in the nucleus. The protons have a positive electric charge, the electrons have a negative electric charge, and the neutrons have no electric charge. If the number of protons and electrons are equal, that atom is electrically neutral. If an atom has more or fewer electrons than protons, then it has an overall negative or positive charge, respectively, and it is called an ion.

The electrons of an atom are attracted to the protons in an atomic nucleus by this electromagnetic force. The protons and neutrons in the nucleus are attracted to each other by a different force, the nuclear force, which is usually stronger than the electromagnetic force repelling the positively charged protons from one another. Under certain circumstances the repelling electromagnetic force becomes stronger than the nuclear force, and nucleons can be ejected from the nucleus, leaving behind a different element: nuclear decay resulting in nuclear transmutation.

The number of protons in the nucleus defines to what chemical element the atom belongs: for example, all copper atoms contain 29 protons. The number of neutrons defines the isotope of the element. The number of electrons influences the magnetic properties of an atom. Atoms can attach to one or more other atoms by chemical bonds to form chemical compounds such as molecules. The ability of atoms to associate and dissociate is responsible for most of the physical changes observed in nature, and is the subject of the discipline of chemistry.

HISTORY OF ATOMIC THEORY

Atoms in Philosophy

The idea that matter is made up of discrete units is a very old idea, appearing in many ancient cultures such as Greece and India. The word "atom" was coined by ancient Greek philosophers. However, these ideas were founded in philosophical and theological reasoning rather than evidence and experimentation. As a result, their views on what atoms look like and how they behave were incorrect. They also could not convince everybody, so atomism was but one of a number of competing theories on the nature of matter. It was not until the 19th century that the idea was embraced and refined by scientists, when the blossoming science of chemistry produced discoveries that only the concept of atoms could explain.

First Evidence-based Theory

In the early 1800s, John Dalton used the concept of atoms to explain why elements always react in ratios of small whole numbers (the law of multiple proportions). For instance, there are two types of tin oxide: one is 88.1% tin and 11.9% oxygen and the other is 78.7% tin and 21.3% oxygen (tin(II) oxide and tin dioxide respectively). This means that 100g of tin will combine either with 13.5g or 27g of oxygen. 13.5 and 27 form a ratio of 1:2, a ratio of small whole numbers. This common pattern in chemistry suggested to Dalton that elements react in whole number multiples of discrete units—in other words, atoms. In the case of tin oxides, one tin atom will combine with either one or two oxygen atoms.

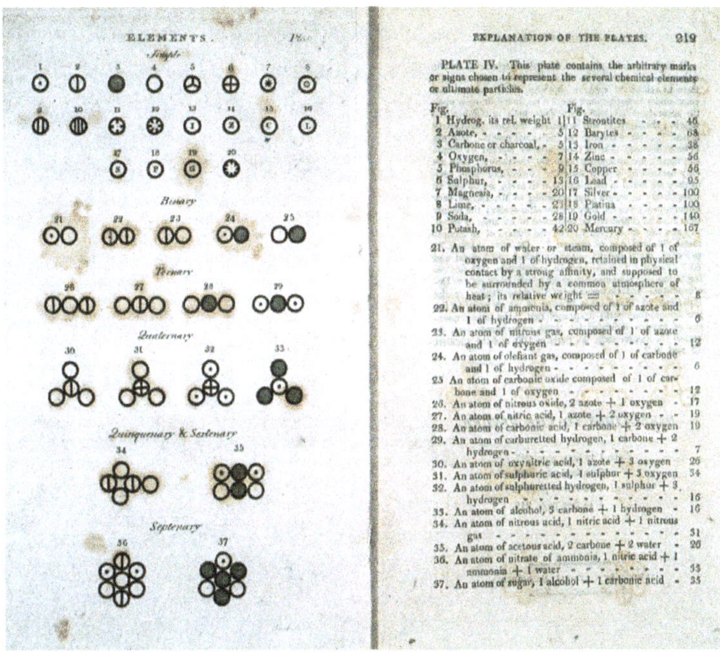

Various atoms and molecules as depicted in John Dalton's A New System of Chemical Philosophy (1808).

Dalton also believed atomic theory could explain why water absorbs different gases in different proportions. For example, he found that water absorbs carbon dioxide far better than it absorbs nitrogen. Dalton hypothesized this was due to the differences between the masses and configurations of the gases' respective particles, and carbon dioxide molecules (CO_2) are heavier and larger than nitrogen molecules (N_2).

Brownian Motion

In 1827, botanist Robert Brown used a microscope to look at dust grains floating in water and discovered that they moved about erratically, a phenomenon that became known as "Brownian motion". This was thought to be caused by water molecules knocking the grains about. In 1905 Albert Einstein proved the reality of these molecules and their motions by producing the first Statistical physics analysis of Brownian motion. French physicist Jean Perrin used Einstein's work to experimentally determine the mass and dimensions of atoms, thereby conclusively verifying Dalton's atomic theory.

Discovery of the Electron

The physicist J. J. Thomson measured the mass of cathode rays, showing they were made of particles, but were around 1800 times lighter than the lightest atom, hydrogen. Therefore, they were not atoms, but a new particle, the first subatomic particle to be discovered, which he originally called "corpuscle" but was later named electron, after particles postulated by George Johnstone Stoney in 1874. He also showed they were identical to particles given off by photoelectric and radioactive materials.

It was quickly recognized that they are the particles that carry electric currents in metal wires, and carry the negative electric charge within atoms. Thomson was given the 1906 Nobel Prize in Physics for this work. Thus he overturned the belief that atoms are the indivisible, ultimate particles of matter. Thomson also incorrectly postulated that the low mass, negatively charged electrons were distributed throughout the atom in a uniform sea of positive charge. This became known as the plum pudding model.

Discovery of the Nucleus

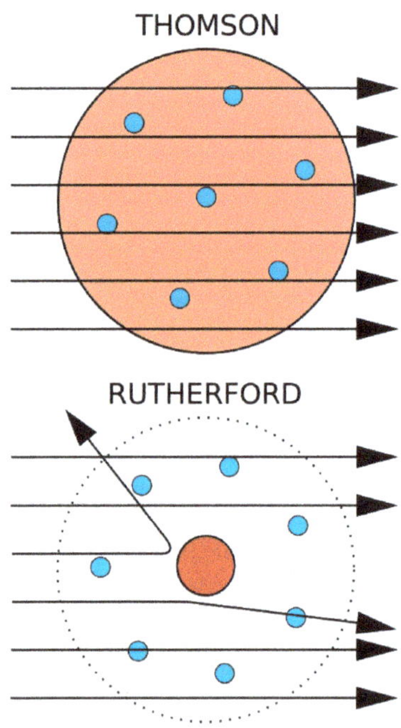

The **Geiger–Marsden experiment**

Top: Expected results: alpha particles passing through the plum pudding model of the atom with negligible deflection.

Bottom: Observed results: a small portion of the particles were deflected by the concentrated positive charge of the nucleus.

In 1909, Hans Geiger and Ernest Marsden, under the direction of Ernest Rutherford, bombarded a metal foil with alpha particles to observe how they scattered. They expected all the alpha particles to pass straight through with little deflection, because Thomson's model said that the charges in the atom are so diffuse that their electric fields could not affect the alpha particles much. However, Geiger and Marsden spotted alpha particles being deflected by angles greater than 90°, which was supposed to be impossible according to Thomson's model. To explain this, Rutherford proposed that the positive charge of the atom is concentrated in a tiny nucleus at the center of the atom. Rutherford compared his findings to firing a 15-inch shell at a sheet of tissue paper and it coming back to hit you.

Discovery of Isotopes

While experimenting with the products of radioactive decay, in 1913 radiochemist Frederick Soddy discovered that there appeared to be more than one type of atom at each position on the periodic table. The term isotope was coined by Margaret Todd as a suitable name for different atoms that belong to the same element. J.J. Thomson created a technique for separating atom types through his work on ionized gases, which subsequently led to the discovery of stable isotopes.

Bohr Model

In 1913 the physicist Niels Bohr proposed a model in which the electrons of an atom were assumed to orbit the nucleus but could only do so in a finite set of orbits, and could jump between these orbits only in discrete changes of energy corresponding to absorption or radiation of a photon. This quantization was used to explain why the electrons orbits are stable (given that normally, charges in acceleration, including circular motion, lose kinetic energy which is emitted as electromagnetic radiation, see synchrotron radiation) and why elements absorb and emit electromagnetic radiation in discrete spectra.

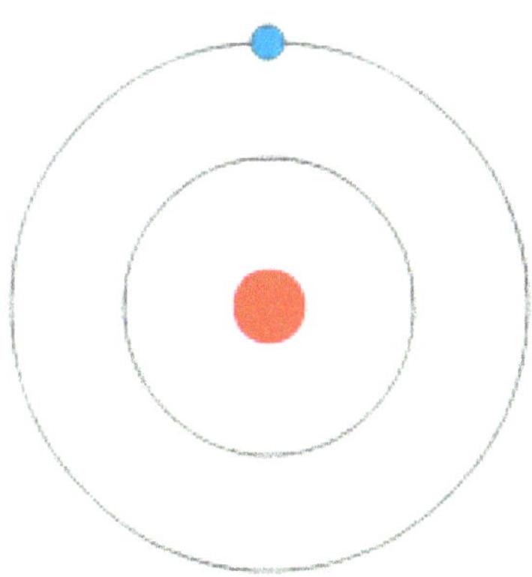

The Bohr model of the atom, with an electron making instantaneous "quantum leaps" from one orbit to another. This model is obsolete.

Later in the same year Henry Moseley provided additional experimental evidence in favor of Niels Bohr's theory. These results refined Ernest Rutherford's and Antonius Van den Broek's model, which proposed that the atom contains in its nucleus a number of positive nuclear charges that is equal to its (atomic) number in the periodic table. Until these experiments, atomic number was not known to be a physical and experimental quantity. That it is equal to the atomic nuclear charge remains the accepted atomic model today.

Chemical Bonding Explained

Chemical bonds between atoms were now explained, by Gilbert Newton Lewis in 1916, as the interactions between their constituent electrons. As the chemical properties of the elements were known to largely repeat themselves according to the periodic law, in 1919 the American chemist Irving Langmuir suggested that this could be explained if the electrons in an atom were connected or clustered in some manner. Groups of electrons were thought to occupy a set of electron shells about the nucleus.

Further Developments in Quantum Physics

The Stern–Gerlach experiment of 1922 provided further evidence of the quantum nature of the atom. When a beam of silver atoms was passed through a specially shaped magnetic field, the beam was split based on the direction of an atom's angular momentum, or spin. As this direction is random, the beam could be expected to spread into a line. Instead, the beam was split into two parts, depending on whether the atomic spin was oriented up or down.

In 1924, Louis de Broglie proposed that all particles behave to an extent like waves. In 1926, Erwin Schrödinger used this idea to develop a mathematical model of the atom that described the electrons as three-dimensional waveforms rather than point particles. A consequence of using waveforms to describe particles is that it is mathematically impossible to obtain precise values for both the position and momentum of a particle at a given point in time; this became known as the uncertainty principle, formulated by Werner Heisenberg in 1926. In this concept, for a given accuracy in measuring a position one could only obtain a range of probable values for momentum, and vice versa. This model was able to explain observations of atomic behavior that previous models could not, such as certain structural and spectral patterns of atoms larger than hydrogen. Thus, the planetary model of the atom was discarded in favor of one that described atomic orbital zones around the nucleus where a given electron is most likely to be observed.

Discovery of the Neutron

The development of the mass spectrometer allowed the mass of atoms to be measured with increased accuracy. The device uses a magnet to bend the trajectory of a beam of ions, and the amount of deflection is determined by the ratio of an atom's mass to its charge. The chemist Francis William Aston used this instrument to show that isotopes had different masses. The atomic mass of these isotopes varied by integer amounts, called the whole number rule. The explanation for these different isotopes awaited the discovery of the neutron, an uncharged particle with a mass similar to the proton, by the physicist James Chadwick in 1932. Isotopes were then explained as elements with the same number of protons, but different numbers of neutrons within the nucleus.

Fission, High-Energy Physics and Condensed Matter

In 1938, the German chemist Otto Hahn, a student of Rutherford, directed neutrons onto uranium atoms expecting to get transuranium elements. Instead, his chemical experiments showed barium as a product. A year later, Lise Meitner and her nephew Otto Frisch verified that Hahn's result were the first experimental nuclear fission. In 1944, Hahn received the Nobel prize in chemistry. Despite Hahn's efforts, the contributions of Meitner and Frisch were not recognized.

In the 1950s, the development of improved particle accelerators and particle detectors allowed scientists to study the impacts of atoms moving at high energies. Neutrons and protons were found to be hadrons, or composites of smaller particles called quarks. The standard model of particle physics was developed that so far has successfully explained the properties of the nucleus in terms of these sub-atomic particles and the forces that govern their interactions.

Subatomic Particles

Though the word atom originally denoted a particle that cannot be cut into smaller particles, in modern scientific usage the atom is composed of various subatomic particles. The constituent particles of an atom are the electron, the proton and the neutron; all three are fermions. However, the hydrogen-1 atom has no neutrons and the hydron ion has no electrons.

The electron is by far the least massive of these particles at 9.11×10^{-31} kg, with a negative electrical charge and a size that is too small to be measured using available techniques. It is the lightest particle with a positive rest mass measured. Under ordinary conditions, electrons are bound to the positively charged nucleus by the attraction created from opposite electric charges. If an atom has more or fewer electrons than its atomic number, then it becomes respectively negatively or positively charged as a whole; a charged atom is called an ion. Electrons have been known since the late 19th century, mostly thanks to J.J. Thomson; see history of subatomic physics for details.

Protons have a positive charge and a mass 1,836 times that of the electron, at 1.6726×10^{-27} kg. The number of protons in an atom is called its atomic number. Ernest Rutherford (1919) observed that nitrogen under alpha-particle bombardment ejects what appeared to be hydrogen nuclei. By 1920 he had accepted that the hydrogen nucleus is a distinct particle within the atom and named it proton.

Neutrons have no electrical charge and have a free mass of 1,839 times the mass of the electron, or 1.6929×10^{-27} kg, the heaviest of the three constituent particles, but it can be reduced by the nuclear binding energy. Neutrons and protons (collectively known as nucleons) have comparable dimensions—on the order of 2.5×10^{-15} m—although the 'surface' of these particles is not sharply defined. The neutron was discovered in 1932 by the English physicist James Chadwick.

In the Standard Model of physics, electrons are truly elementary particles with no internal structure. However, both protons and neutrons are composite particles composed of elementary particles called quarks. There are two types of quarks in atoms, each having a fractional electric charge. Protons are composed of two up quarks (each with charge +2/3) and one down quark (with a charge of −1/3). Neutrons consist of one up quark and two down quarks. This distinction accounts for the difference in mass and charge between the two particles.

The quarks are held together by the strong interaction (or strong force), which is mediated by gluons. The protons and neutrons, in turn, are held to each other in the nucleus by the nuclear force, which is a residuum of the strong force that has somewhat different range-properties (see the article on the nuclear force for more). The gluon is a member of the family of gauge bosons, which are elementary particles that mediate physical forces.

Nucleus

All the bound protons and neutrons in an atom make up a tiny atomic nucleus, and are collectively called nucleons. The radius of a nucleus is approximately equal to 1.07 $\sqrt[3]{A}$ fm, where A is the total number of nucleons. This is much smaller than the radius of the atom, which is on the order of 10^5 fm. The nucleons are bound together by a short-ranged attractive potential called the residual strong force. At distances smaller than 2.5 fm this force is much more powerful than the electrostatic force that causes positively charged protons to repel each other.

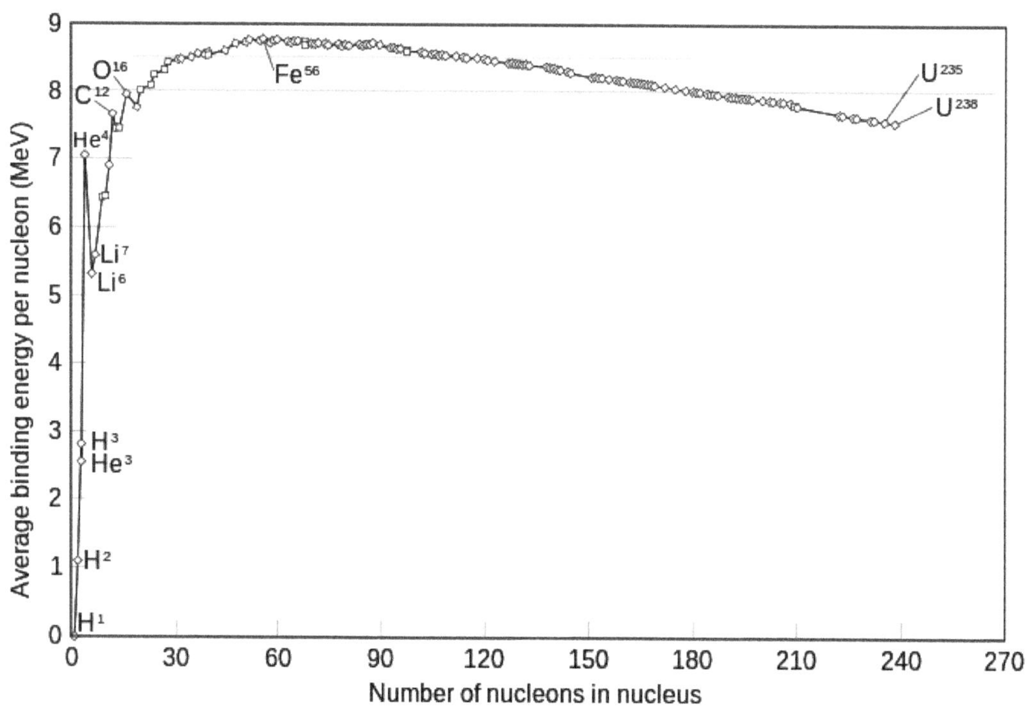

The binding energy needed for a nucleon to escape the nucleus, for various isotopes.

Atoms of the same element have the same number of protons, called the atomic number. Within a single element, the number of neutrons may vary, determining the isotope of that element. The total number of protons and neutrons determine the nuclide. The number of neutrons relative to the protons determines the stability of the nucleus, with certain isotopes undergoing radioactive decay.

The proton, the electron, and the neutron are classified as fermions. Fermions obey the Pauli exclusion principle which prohibits identical fermions, such as multiple protons, from occupying the same quantum state at the same time. Thus, every proton in the nucleus must occupy a quantum state different from all other protons, and the same applies to all neutrons of the nucleus and to all electrons of the electron cloud. However, a proton and a neutron are allowed to occupy the same quantum state.

For atoms with low atomic numbers, a nucleus that has more neutrons than protons tends to drop to a lower energy state through radioactive decay so that the neutron–proton ratio is closer to one. However, as the atomic number increases, a higher proportion of neutrons is required to offset the mutual repulsion of the protons. Thus, there are no stable nuclei with equal proton and neutron numbers above atomic number $Z = 20$ (calcium) and as Z increases, the neutron–proton ratio of stable isotopes increases. The stable isotope with the highest proton–neutron ratio is lead-208 (about 1.5).

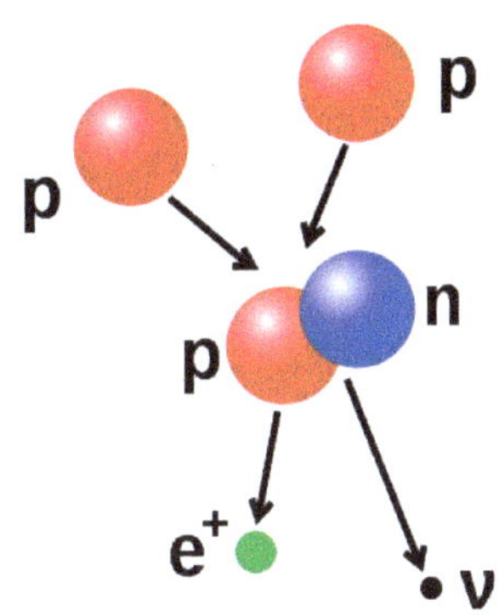

Illustration of a nuclear fusion process that forms a deuterium nucleus, consisting of a proton and a neutron, from two protons. A positron (e^+)—an antimatter electron—is emitted along with an electron neutrino.

The number of protons and neutrons in the atomic nucleus can be modified, although this can require very high energies because of the strong force. Nuclear fusion occurs when multiple atomic particles join to form a heavier nucleus, such as through the energetic collision of two nuclei. For example, at the core of the Sun protons require energies of 3–10 keV to overcome their mutual repulsion—the coulomb barrier—and fuse together into a single nucleus. Nuclear fission is the opposite process, causing a nucleus to split into two smaller nuclei—usually through radioactive decay. The nucleus can also be modified through bombardment by high energy subatomic particles or photons. If this modifies the number of protons in a nucleus, the atom changes to a different chemical element.

If the mass of the nucleus following a fusion reaction is less than the sum of the masses of the separate particles, then the difference between these two values can be emitted as a type of usable energy (such as a gamma ray, or the kinetic energy of a beta particle), as described by Albert Einstein's mass–energy equivalence formula, $E = mc^2$, where m is the mass loss and c is the speed of light. This deficit is part of the binding energy of the new nucleus, and it is the non-recoverable loss of the energy that causes the fused particles to remain together in a state that requires this energy to separate.

The fusion of two nuclei that create larger nuclei with lower atomic numbers than iron and nickel—a total nucleon number of about 60—is usually an exothermic process that releases more energy than is required to bring them together. It is this energy-releasing process that makes nuclear fusion in stars a self-sustaining reaction. For heavier nuclei, the binding energy per nucleon in the nucleus begins to decrease. That means fusion processes producing nuclei that have atomic numbers higher than about 26, and atomic masses higher than about 60, is an endothermic process. These more massive nuclei cannot undergo an energy-producing fusion reaction that can sustain the hydrostatic equilibrium of a star.

Electron Cloud

The electrons in an atom are attracted to the protons in the nucleus by the electromagnetic force. This force binds the electrons inside an electrostatic potential well surrounding the smaller nucleus, which means that an external source of energy is needed for the electron to escape. The closer an electron is to the nucleus, the greater the attractive force. Hence electrons bound near the center of the potential well require more energy to escape than those at greater separations.

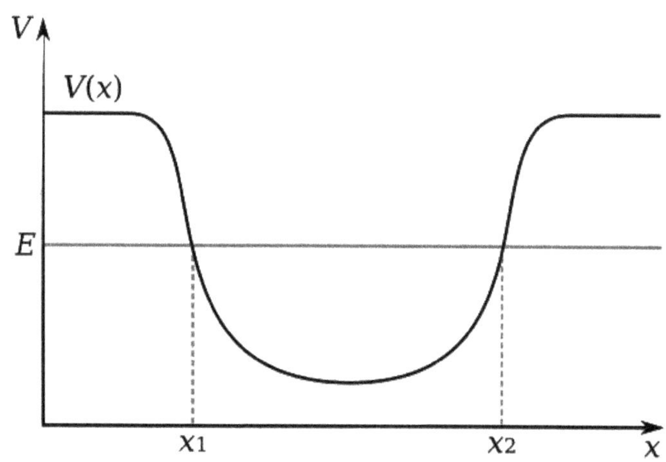

A potential well, showing, according to classical mechanics, the minimum energy V(x) needed to reach each position x. Classically, a particle with energy E is constrained to a range of positions between x_1 and x_2.

Electrons, like other particles, have properties of both a particle and a wave. The electron cloud is a region inside the potential well where each electron forms a type of three-dimensional standing wave—a wave form that does not move relative to the nucleus. This behavior is defined by an atomic orbital, a mathematical function that characterises the probability that an electron appears to be at a particular location when its position is measured. Only a discrete (or quantized) set of these orbitals exist around the nucleus, as other possible wave patterns rapidly decay into a more stable form. Orbitals can have one or more ring or node structures, and differ from each other in size, shape and orientation.

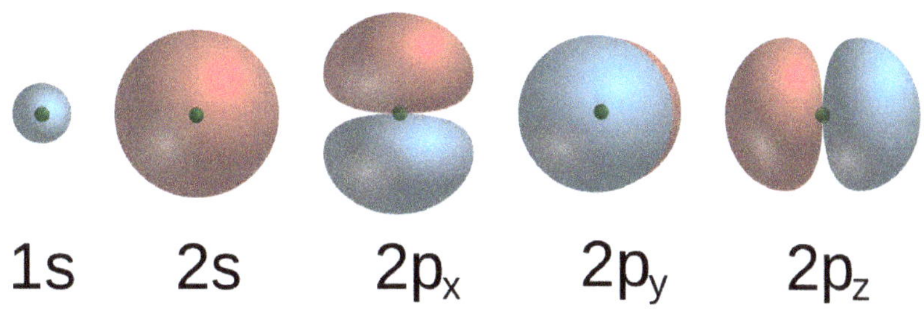

Wave functions of the first five atomic orbitals. The three 2p orbitals each display a single angular node that has an orientation and a minimum at the center.

Each atomic orbital corresponds to a particular energy level of the electron. The electron can change its state to a higher energy level by absorbing a photon with sufficient energy to boost it into the new quantum state. Likewise, through spontaneous emission, an electron in a higher energy state can drop to a lower energy state while radiating the excess energy as a photon. These characteristic energy values, defined by the differences in the energies of the quantum states, are responsible for atomic spectral lines.

The amount of energy needed to remove or add an electron—the electron binding energy—is far less than the binding energy of nucleons. For example, it requires only 13.6 eV to strip a ground-state electron from a hydrogen atom, compared to 2.23 million eV for splitting a deuterium nucleus. Atoms are electrically neutral if they have an equal number of protons and electrons. Atoms that have either a deficit or a surplus of electrons are called ions. Electrons that are farthest from the nucleus may be transferred to other nearby atoms or shared between atoms. By this mechanism, atoms are able to bond into molecules and other types of chemical compounds like ionic and covalent network crystals.

Properties

Nuclear Properties

By definition, any two atoms with an identical number of protons in their nuclei belong to the same chemical element. Atoms with equal numbers of protons but a different number of neutrons are different isotopes of the same element. For example, all hydrogen atoms admit exactly one proton, but isotopes exist with no neutrons (hydrogen-1, by far the most common form, also called protium), one neutron (deuterium), two neutrons (tritium) and more than two neutrons. The known elements form a set of atomic numbers, from the single proton element hydrogen up to the 118-proton element oganesson. All known isotopes of elements with atomic numbers greater than 82 are radioactive, although the radioactivity of element 83 (bismuth) is so slight as to be practically negligible.

About 339 nuclides occur naturally on Earth, of which 254 (about 75%) have not been observed to decay, and are referred to as "stable isotopes". However, only 90 of these nuclides are stable to all decay, even in theory. Another 164 (bringing the total to 254) have not been observed to decay, even though in theory it is energetically possible. These are also formally classified as "stable". An additional 34 radioactive nuclides have half-lives longer than 80 million years, and are long-lived enough to be present from the birth of the solar system. This collection of 288 nuclides are known as primordial nuclides. Finally, an additional 51 short-lived nuclides are known to occur naturally, as daughter products of primordial nuclide decay (such as radium from uranium), or else as products of natural energetic processes on Earth, such as cosmic ray bombardment (for example, carbon-14).

For 80 of the chemical elements, at least one stable isotope exists. As a rule, there is only a handful of stable isotopes for each of these elements, the average being 3.2 stable isotopes per element. Twenty-six elements have only a single stable isotope, while the largest number of stable isotopes observed for any element is ten, for the element tin. Elements 43, 61, and all elements numbered 83 or higher have no stable isotopes.

Stability of isotopes is affected by the ratio of protons to neutrons, and also by the presence of certain "magic numbers" of neutrons or protons that represent closed and filled quantum shells. These quantum shells correspond to a set of energy levels within the shell model of the nucleus; filled shells, such as the filled shell of 50 protons for tin, confers unusual stability on the nuclide. Of the 254 known stable nuclides, only four have both an odd number of protons and odd number of neutrons: hydrogen-2 (deuterium), lithium-6, boron-10 and nitrogen-14. Also, only four naturally occurring, radioactive odd–odd nuclides have a half-life over a billion years: potassium-40, vanadium-50, lanthanum-138 and tantalum-180m. Most odd–odd nuclei are highly unstable with respect to beta decay, because the decay products are even–even, and are therefore more strongly bound, due to nuclear pairing effects.

Mass

The large majority of an atom's mass comes from the protons and neutrons that make it up. The total number of these particles (called "nucleons") in a given atom is called the mass number. It is a positive integer and dimensionless (instead of having dimension of mass), because it expresses a count. An example of use of a mass number is "carbon-12," which has 12 nucleons (six protons and six neutrons), another example with unequal number of protons and neutrons, such as carbon-14, so that students will not think that the total number of nucleons is half protons, half neutrons. another example with unequal number of protons and neutrons, such as carbon-14, so that students will not think that the total number of nucleons is half protons, half neutrons.

The actual mass of an atom at rest is often expressed using the unified atomic mass unit (u), also called dalton (Da). This unit is defined as a twelfth of the mass of a free neutral atom of carbon-12, which is approximately 1.66×10^{-27} kg. Hydrogen-1 (the lightest isotope of hydrogen which is also the nuclide with the lowest mass) has an atomic weight of 1.007825 u. The value of this number is called the atomic mass. A given atom has an atomic mass approximately equal (within 1%) to its mass number times the atomic mass unit (for example the mass of a nitrogen-14 is roughly 14 u). However, this number will not be exactly an integer except in the case of carbon-12 (see below). The heaviest stable atom is lead-208, with a mass of 207.9766521 u.

As even the most massive atoms are far too light to work with directly, chemists instead use the unit of moles. One mole of atoms of any element always has the same number of atoms (about 6.022×10^{23}). This number was chosen so that if an element has an atomic mass of 1 u, a mole of atoms of that element has a mass close to one gram. Because of the definition of the unified atomic mass unit, each carbon-12 atom has an atomic mass of exactly 12 u, and so a mole of carbon-12 atoms weighs exactly 0.012 kg.

Shape and Size

Atoms lack a well-defined outer boundary, so their dimensions are usually described in terms of an atomic radius. This is a measure of the distance out to which the electron cloud extends from the nucleus. However, this assumes the atom to exhibit a spherical shape, which is only obeyed for atoms in vacuum or free space. Atomic radii may be derived from the distances between two nuclei when the two atoms are joined in a chemical bond. The radius varies with the location of an atom on the atomic chart, the type of chemical bond, the number of neighboring atoms (coordination number) and a quantum mechanical property known as spin. On the periodic table of the elements, atom size tends to increase when moving down columns, but decrease when moving across rows (left to right). Consequently, the smallest atom is helium with a radius of 32 pm, while one of the largest is caesium at 225 pm.

When subjected to external forces, like electrical fields, the shape of an atom may deviate from spherical symmetry. The deformation depends on the field magnitude and the orbital type of outer shell electrons, as shown by group-theoretical considerations. Aspherical deviations might be elicited for instance in crystals, where large crystal-electrical fields may occur at low-symmetry lattice sites. Significant ellipsoidal deformations have recently been shown to occur for sulfur ions and chalcogen ions in pyrite-type compounds.

Atomic dimensions are thousands of times smaller than the wavelengths of light (400–700 nm) so they cannot be viewed using an optical microscope. However, individual atoms can be observed using a scanning tunneling microscope. To visualize the minuteness of the atom, consider that a typical human hair is about 1 million carbon atoms in width. A single drop of water contains about 2 sextillion (2×10^{21}) atoms of oxygen, and twice the number of hydrogen atoms. A single carat diamond with a mass of 2×10^{-4} kg contains about 10 sextillion (10^{22}) atoms of carbon. If an apple were magnified to the size of the Earth, then the atoms in the apple would be approximately the size of the original apple.

Radioactive Decay

Every element has one or more isotopes that have unstable nuclei that are subject to radioactive decay, causing the nucleus to emit particles or electromagnetic radiation.

Radioactivity can occur when the radius of a nucleus is large compared with the radius of the strong force, which only acts over distances on the order of 1 fm.

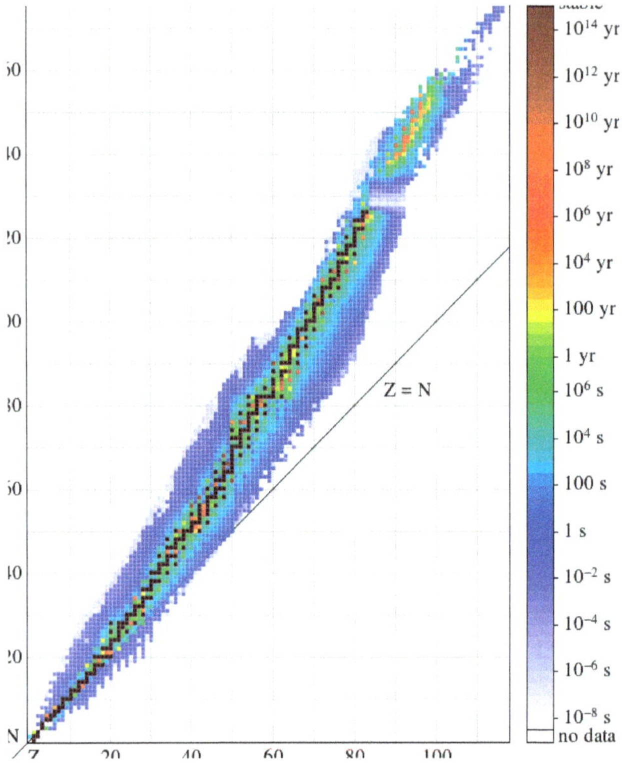

This diagram shows the half-life ($T_{1/2}$) of various isotopes with Z protons and N neutrons.

The most common forms of radioactive decay are:

- Alpha decay: this process is caused when the nucleus emits an alpha particle, which is a helium nucleus consisting of two protons and two neutrons. The result of the emission is a new element with a lower atomic number.
- Beta decay (and electron capture): these processes are regulated by the weak force, and result from a transformation of a neutron into a proton, or a proton into a neutron. The neutron to proton transition is accompanied by the emission of an electron and an antineutrino, while proton to neutron transition (except in electron capture) causes the emission of a positron and a neutrino. The electron or positron emissions are called beta particles. Beta decay either increases or decreases the atomic number of the nucleus by one. Electron capture is more common than positron emission, because it requires less energy. In this type of decay, an electron is absorbed by the nucleus, rather than a positron emitted from the nucleus. A neutrino is still emitted in this process, and a proton changes to a neutron.
- Gamma decay: this process results from a change in the energy level of the nucleus to a lower state, resulting in the emission of electromagnetic radiation. The excited state of a nucleus which results in gamma emission usually occurs

following the emission of an alpha or a beta particle. Thus, gamma decay usually follows alpha or beta decay.

Other more rare types of radioactive decay include ejection of neutrons or protons or clusters of nucleons from a nucleus, or more than one beta particle. An analog of gamma emission which allows excited nuclei to lose energy in a different way, is internal conversion— a process that produces high-speed electrons that are not beta rays, followed by production of high-energy photons that are not gamma rays. A few large nuclei explode into two or more charged fragments of varying masses plus several neutrons, in a decay called spontaneous nuclear fission.

Each radioactive isotope has a characteristic decay time period—the half-life—that is determined by the amount of time needed for half of a sample to decay. This is an exponential decay process that steadily decreases the proportion of the remaining isotope by 50% every half-life. Hence after two half-lives have passed only 25% of the isotope is present, and so forth.

Magnetic Moment

Elementary particles possess an intrinsic quantum mechanical property known as spin. This is analogous to the angular momentum of an object that is spinning around its center of mass, although strictly speaking these particles are believed to be point-like and cannot be said to be rotating. Spin is measured in units of the reduced Planck constant (ħ), with electrons, protons and neutrons all having spin ½ ħ, or "spin-½". In an atom, electrons in motion around the nucleus possess orbital angular momentum in addition to their spin, while the nucleus itself possesses angular momentum due to its nuclear spin.

The magnetic field produced by an atom—its magnetic moment—is determined by these various forms of angular momentum, just as a rotating charged object classically produces a magnetic field. However, the most dominant contribution comes from electron spin. Due to the nature of electrons to obey the Pauli exclusion principle, in which no two electrons may be found in the same quantum state, bound electrons pair up with each other, with one member of each pair in a spin up state and the other in the opposite, spin down state. Thus these spins cancel each other out, reducing the total magnetic dipole moment to zero in some atoms with even number of electrons.

In ferromagnetic elements such as iron, cobalt and nickel, an odd number of electrons leads to an unpaired electron and a net overall magnetic moment. The orbitals of neighboring atoms overlap and a lower energy state is achieved when the spins of unpaired electrons are aligned with each other, a spontaneous process known as an exchange interaction. When the magnetic moments of ferromagnetic atoms are lined up, the material can produce a measurable macroscopic field. Paramagnetic materials have atoms with magnetic moments that line up in random directions when no magnetic field is present, but the magnetic moments of the individual atoms line up in the presence of a field.

The nucleus of an atom will have no spin when it has even numbers of both neutrons and protons, but for other cases of odd numbers, the nucleus may have a spin. Normally nuclei with spin are aligned in random directions because of thermal equilibrium. However, for certain elements (such as xenon-129) it is possible to polarize a significant proportion of the nuclear spin states so that they are aligned in the same direction—a condition called hyperpolarization. This has important applications in magnetic resonance imaging.

Energy Levels

The potential energy of an electron in an atom is negative, its dependence of its position reaches the minimum (the most absolute value) inside the nucleus, and vanishes when the distance from the nucleus goes to infinity, roughly in an inverse proportion to the distance. In the quantum-mechanical model, a bound electron can only occupy a set of states centered on the nucleus, and each state corresponds to a specific energy level; see time-independent Schrödinger equation for theoretical explanation. An energy level can be measured by the amount of energy needed to unbind the electron from the atom, and is usually given in units of electronvolts (eV). The lowest energy state of a bound electron is called the ground state, i.e. stationary state, while an electron transition to a higher level results in an excited state. The electron's energy raises when n increases because the (average) distance to the nucleus increases. Dependence of the energy on ℓ is caused not by electrostatic potential of the nucleus, but by interaction between electrons.

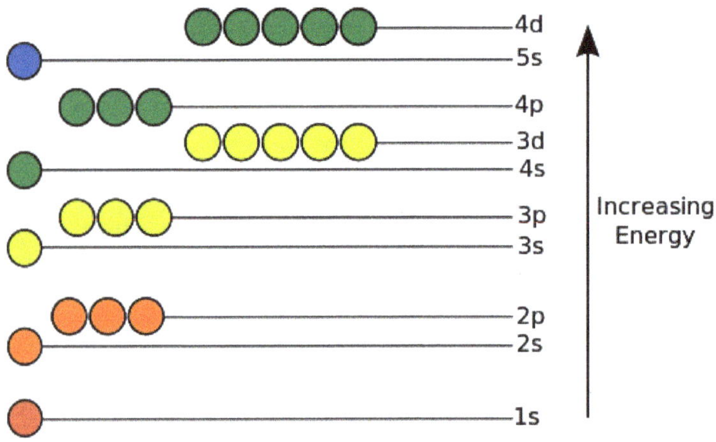

These electron's energy levels (not to scale) are sufficient for ground states of atoms up to cadmium ($5s^2$ $4d^{10}$) inclusively. Do not forget that even the top of the diagram is lower than an unbound electron state.

For an electron to transition between two different states, e.g. grounded state to first excited level (ionization), it must absorb or emit a photon at an energy matching the difference in the potential energy of those levels, according to Niels Bohr model, what can be precisely calculated by the Schrödinger equation. Electrons jump between orbitals in a particle-like fashion. For example, if a single photon strikes the electrons, only a single electron changes states in response to the photon; see Electron properties.

The energy of an emitted photon is proportional to its frequency, so these specific energy levels appear as distinct bands in the electromagnetic spectrum. Each element has a characteristic spectrum that can depend on the nuclear charge, subshells filled by electrons, the electromagnetic interactions between the electrons and other factors.

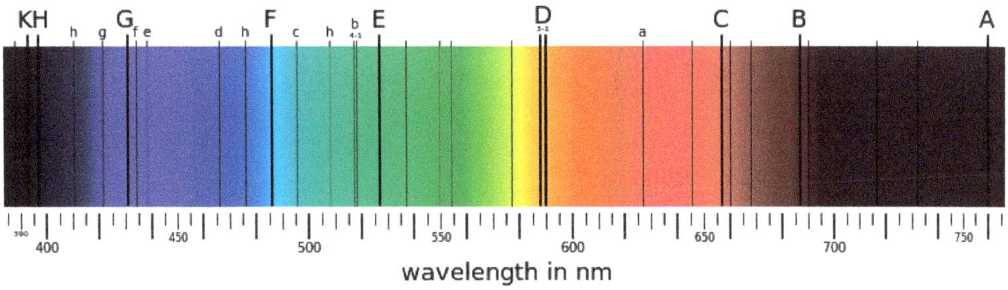

An example of absorption lines in a spectrum

When a continuous spectrum of energy is passed through a gas or plasma, some of the photons are absorbed by atoms, causing electrons to change their energy level. Those excited electrons that remain bound to their atom spontaneously emit this energy as a photon, traveling in a random direction, and so drop back to lower energy levels. Thus the atoms behave like a filter that forms a series of dark absorption bands in the energy output. (An observer viewing the atoms from a view that does not include the continuous spectrum in the background, instead sees a series of emission lines from the photons emitted by the atoms.) Spectroscopic measurements of the strength and width of atomic spectral lines allow the composition and physical properties of a substance to be determined.

Close examination of the spectral lines reveals that some display a fine structure splitting. This occurs because of spin–orbit coupling, which is an interaction between the spin and motion of the outermost electron. When an atom is in an external magnetic field, spectral lines become split into three or more components; a phenomenon called the Zeeman effect. This is caused by the interaction of the magnetic field with the magnetic moment of the atom and its electrons. Some atoms can have multiple electron configurations with the same energy level, which thus appear as a single spectral line. The interaction of the magnetic field with the atom shifts these electron configurations to slightly different energy levels, resulting in multiple spectral lines. The presence of an external electric field can cause a comparable splitting and shifting of spectral lines by modifying the electron energy levels, a phenomenon called the Stark effect.

If a bound electron is in an excited state, an interacting photon with the proper energy can cause stimulated emission of a photon with a matching energy level. For this to occur, the electron must drop to a lower energy state that has an energy difference matching the energy of the interacting photon. The emitted photon and the interacting photon then move off in parallel and with matching phases. That is, the wave patterns of the two photons are synchronized. This physical property is used to make lasers, which can emit a coherent beam of light energy in a narrow frequency band.

Valence (chemistry) and Chemical bond

Valency is the combining power of an element. It is equal to number of hydrogen atoms that atom can combine or displace in forming compounds. The outermost electron shell of an atom in its uncombined state is known as the valence shell, and the electrons in that shell are called valence electrons. The number of valence electrons determines the bonding behavior with other atoms. Atoms tend to chemically react with each other in a manner that fills (or empties) their outer valence shells. For example, a transfer of a single electron between atoms is a useful approximation for bonds that form between atoms with one-electron more than a filled shell, and others that are one-electron short of a full shell, such as occurs in the compound sodium chloride and other chemical ionic salts. However, many elements display multiple valences, or tendencies to share differing numbers of electrons in different compounds. Thus, chemical bonding between these elements takes many forms of electron-sharing that are more than simple electron transfers. Examples include the element carbon and the organic compounds.

The chemical elements are often displayed in a periodic table that is laid out to display recurring chemical properties, and elements with the same number of valence electrons form a group that is aligned in the same column of the table. (The horizontal rows correspond to the filling of a quantum shell of electrons.) The elements at the far right of the table have their outer shell completely filled with electrons, which results in chemically inert elements known as the noble gases.

States

Quantities of atoms are found in different states of matter that depend on the physical conditions, such as temperature and pressure. By varying the conditions, materials can transition between solids, liquids, gases and plasmas. Within a state, a material can also exist in different allotropes. An example of this is solid carbon, which can exist as graphite or diamond. Gaseous allotropes exist as well, such as dioxygen and ozone.

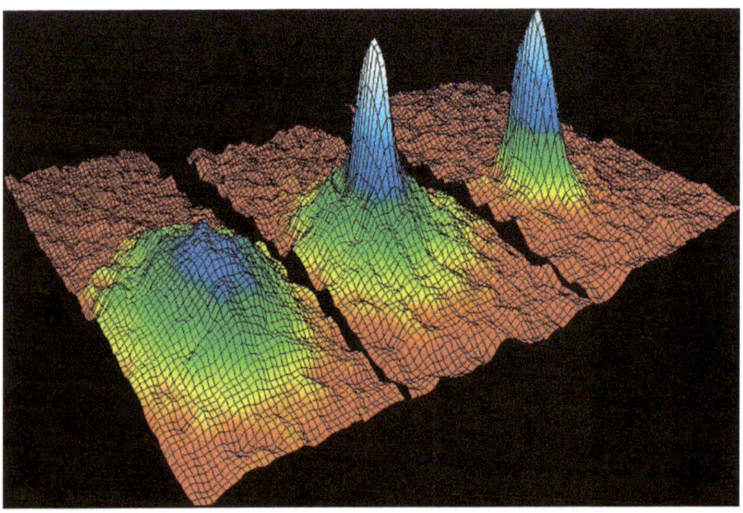

Snapshots illustrating the formation of a Bose–Einstein condensate.

At temperatures close to absolute zero, atoms can form a Bose–Einstein condensate, at which point quantum mechanical effects, which are normally only observed at the atomic scale, become apparent on a macroscopic scale. This super-cooled collection of atoms then behaves as a single super atom, which may allow fundamental checks of quantum mechanical behavior.

Identification

The scanning tunneling microscope is a device for viewing surfaces at the atomic level. It uses the quantum tunneling phenomenon, which allows particles to pass through a barrier that would normally be insurmountable. Electrons tunnel through the vacuum between two planar metal electrodes, on each of which is an adsorbed atom, providing a tunneling-current density that can be measured. Scanning one atom (taken as the tip) as it moves past the other (the sample) permits plotting of tip displacement versus lateral separation for a constant current. The calculation shows the extent to which scanning-tunneling-microscope images of an individual atom are visible. It confirms that for low bias, the microscope images the space-averaged dimensions of the electron orbitals across closely packed energy levels—the Fermi level local density of states.

Scanning tunneling microscope image showing the individual atoms making up this gold (100) surface. The surface atoms deviate from the bulk crystal structure and arrange in columns several atoms wide with pits between them.

An atom can be ionized by removing one of its electrons. The electric charge causes the trajectory of an atom to bend when it passes through a magnetic field. The radius by which the trajectory of a moving ion is turned by the magnetic field is determined by the mass of the atom. The mass spectrometer uses this principle to measure the mass-to-charge ratio of ions. If a sample contains multiple isotopes, the mass spectrometer can determine the proportion of each isotope in the sample by measuring the intensity of the different beams of ions. Techniques to vaporize atoms include inductively coupled plasma atomic emission spectroscopy and inductively coupled plasma mass spectrometry, both of which use a plasma to vaporize samples for analysis.

A more area-selective method is electron energy loss spectroscopy, which measures the energy loss of an electron beam within a transmission electron microscope when it interacts with a portion of a sample. The atom-probe tomograph has sub-nanometer resolution in 3-D and can chemically identify individual atoms using time-of-flight mass spectrometry.

Spectra of excited states can be used to analyze the atomic composition of distant stars. Specific light wavelengths contained in the observed light from stars can be separated out and related to the quantized transitions in free gas atoms. These colors can be replicated using a gas-discharge lamp containing the same element. Helium was discovered in this way in the spectrum of the Sun 23 years before it was found on Earth.

Origin and current state

Atoms form about 4% of the total energy density of the observable Universe, with an average density of about 0.25 atoms/m^3. Within a galaxy such as the Milky Way, atoms have a much higher concentration, with the density of matter in the interstellar medium (ISM) ranging from 10^5 to 10^9 atoms/m^3. The Sun is believed to be inside the Local Bubble, a region of highly ionized gas, so the density in the solar neighborhood is only about 10^3 atoms/m^3. Stars form from dense clouds in the ISM, and the evolutionary processes of stars result in the steady enrichment of the ISM with elements more massive than hydrogen and helium. Up to 95% of the Milky Way's atoms are concentrated inside stars and the total mass of atoms forms about 10% of the mass of the galaxy. (The remainder of the mass is an unknown dark matter.)

Formation

Electrons are thought to exist in the Universe since early stages of the Big Bang. Atomic nuclei forms in nucleosynthesis reactions. In about three minutes Big Bang nucleosynthesis produced most of the helium, lithium, and deuterium in the Universe, and perhaps some of the beryllium and boron.

Ubiquitousness and stability of atoms relies on their binding energy, which means that an atom has a lower energy than an unbound system of the nucleus and electrons. Where the temperature is much higher than ionization potential, the matter exists in the form of plasma—a gas of positively charged ions (possibly, bare nuclei) and electrons. When the temperature drops below the ionization potential, atoms become statistically favorable. Atoms (complete with bound electrons) became to dominate over charged particles 380,000 years after the Big Bang—an epoch called recombination, when the expanding Universe cooled enough to allow electrons to become attached to nuclei.

Since the Big Bang, which produced no carbon or heavier elements, atomic nuclei have been combined in stars through the process of nuclear fusion to produce more of the element helium, and (via the triple alpha process) the sequence of elements from carbon up to iron; see stellar nucleosynthesis for details.

Isotopes such as lithium-6, as well as some beryllium and boron are generated in space through cosmic ray spallation. This occurs when a high-energy proton strikes an atomic nucleus, causing large numbers of nucleons to be ejected.

Elements heavier than iron were produced in supernovae through the r-process and in AGB stars through the s-process, both of which involve the capture of neutrons by atomic nuclei. Elements such as lead formed largely through the radioactive decay of heavier elements.

Earth

Most of the atoms that make up the Earth and its inhabitants were present in their current form in the nebula that collapsed out of a molecular cloud to form the Solar System. The rest are the result of radioactive decay, and their relative proportion can be used to determine the age of the Earth through radiometric dating. Most of the helium in the crust of the Earth (about 99% of the helium from gas wells, as shown by its lower abundance of helium-3) is a product of alpha decay.

There are a few trace atoms on Earth that were not present at the beginning (i.e., not "primordial"), nor are results of radioactive decay. Carbon-14 is continuously generated by cosmic rays in the atmosphere. Some atoms on Earth have been artificially generated either deliberately or as by-products of nuclear reactors or explosions. Of the transuranic elements—those with atomic numbers greater than 92—only plutonium and neptunium occur naturally on Earth. Transuranic elements have radioactive lifetimes shorter than the current age of the Earth and thus identifiable quantities of these elements have long since decayed, with the exception of traces of plutonium-244 possibly deposited by cosmic dust. Natural deposits of plutonium and neptunium are produced by neutron capture in uranium ore.

The Earth contains approximately 1.33×10^{50} atoms. Although small numbers of independent atoms of noble gases exist, such as argon, neon, and helium, 99% of the atmosphere is bound in the form of molecules, including carbon dioxide and diatomic oxygen and nitrogen. At the surface of the Earth, an overwhelming majority of atoms combine to form various compounds, including water, salt, silicates and oxides. Atoms can also combine to create materials that do not consist of discrete molecules, including crystals and liquid or solid metals. This atomic matter forms networked arrangements that lack the particular type of small-scale interrupted order associated with molecular matter.

Rare and Theoretical Forms

Superheavy Elements

While isotopes with atomic numbers higher than lead (82) are known to be radioactive, an "island of stability" has been proposed for some elements with atomic numbers above 103. These superheavy elements may have a nucleus that is relatively stable against radioactive decay. The most likely candidate for a stable superheavy atom, unbihexium, has 126 protons and 184 neutrons.

Exotic Matter

Each particle of matter has a corresponding antimatter particle with the opposite electrical charge. Thus, the positron is a positively charged antielectron and the antiproton is a negatively charged equivalent of a proton. When a matter and corresponding antimatter particle meet, they annihilate each other. Because of this, along with an imbalance between the number of matter and antimatter particles, the latter are rare in the universe. The first causes of this imbalance are not yet fully understood, although theories of baryogenesis may offer an explanation. As a result, no antimatter atoms have been discovered in nature. However, in 1996 the antimatter counterpart of the hydrogen atom (antihydrogen) was synthesized at the CERN laboratory in Geneva.

Other exotic atoms have been created by replacing one of the protons, neutrons or electrons with other particles that have the same charge. For example, an electron can be replaced by a more massive muon, forming a muonic atom. These types of atoms can be used to test the fundamental predictions of physics.

ATOMIC NUCLEUS

The atomic nucleus is the small, dense region consisting of protons and neutrons at the center of an atom, discovered in 1911 by Ernest Rutherford based on the 1909 Geiger–Marsden gold foil experiment. After the discovery of the neutron in 1932, models for a nucleus composed of protons and neutrons were quickly developed by Dmitri Ivanenko and Werner Heisenberg. Almost all of the mass of an atom is located in the nucleus, with a very small contribution from the electron cloud. Protons and neutrons are bound together to form a nucleus by the nuclear force.

The diameter of the nucleus is in the range of 1.75 fm (1.75×10^{-15} m) for hydrogen (the diameter of a single proton) to about 15 fm for the heaviest atoms, such as uranium. These dimensions are much smaller than the diameter of the atom itself (nucleus + electron cloud), by a factor of about 23,000 (uranium) to about 145,000 (hydrogen).

The branch of physics concerned with the study and understanding of the atomic nucleus, including its composition and the forces which bind it together, is called nuclear physics.

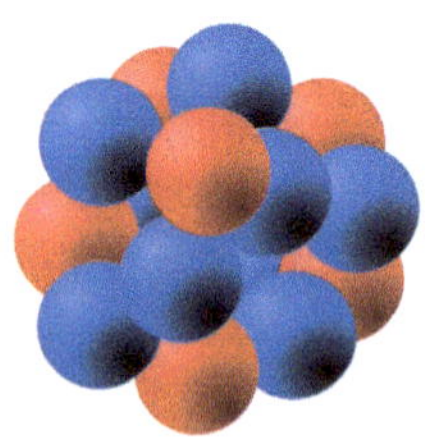

A model of the atomic nucleus showing it as a compact bundle of the two types of nucleons: protons (red) and neutrons (blue). In this diagram, protons and neutrons look like little balls stuck together, but an actual nucleus (as understood by modern nuclear physics) cannot be explained like this, but only by using quantum mechanics. In a nucleus which occupies a certain energy level (for example, the ground state), each nucleon can be said to occupy a range of locations.

History

The nucleus was discovered in 1911, as a result of Ernest Rutherford's efforts to test Thomson's "plum pudding model" of the atom. The electron had already been discovered earlier by J.J. Thomson himself. Knowing that atoms are electrically neutral, Thomson postulated that there must be a positive charge as well. In his plum pudding model, Thomson suggested that an atom consisted of negative electrons randomly scattered within a sphere of positive charge. Ernest Rutherford later devised an experiment with his research partner Hans Geiger and with help of Ernest Marsden, that involved the deflection of alpha particles (helium nuclei) directed at a thin sheet of metal foil. He reasoned that if Thomson's model were correct, the positively charged alpha particles would easily pass through the foil with very little deviation in their paths, as the foil should act as electrically neutral if the negative and positive charges are so intimately mixed as to make it appear neutral. To his surprise, many of the particles were deflected at very large angles. Because the mass of an alpha particle is about 8000 times that of an electron, it became apparent that a very strong force must be present if it could deflect the massive and fast moving alpha particles. He realized that the plum pudding model could not be accurate and that the deflections of the alpha particles could only be explained if the positive and negative charges were separated from each other and that the mass of the atom was a concentrated point of positive charge. This justified the idea of a nuclear atom with a dense center of positive charge and mass.

Etymology

The term nucleus is from the Latin word nucleus, a diminutive of nux ("nut"), meaning the kernel (i.e., the "small nut") inside a watery type of fruit (like a peach). In 1844, Michael Faraday used the term to refer to the "central point of an atom". The modern atomic meaning was proposed by Ernest Rutherford in 1912. The adoption of the term "nucleus" to atomic theory, however, was not immediate. In 1916, for example, Gilbert N. Lewis stated, in his famous article The Atom and the Molecule, that "the atom is composed of the kernel and an outer atom or shell"

Nuclear Makeup

The nucleus of an atom consists of neutrons and protons, which in turn are the manifestation of more elementary particles, called quarks, that are held in association by the nuclear strong force in certain stable combinations of hadrons, called baryons. The nuclear strong force extends far enough from each baryon so as to bind the neutrons and protons together against the repulsive electrical force between the positively charged protons. The nuclear strong force has a very short range, and essentially drops to zero just beyond the edge of the nucleus. The collective action of the positively charged nucleus is to hold the electrically negative charged electrons in their orbits about the nucleus. The collection of negatively charged electrons orbiting the nucleus display an affinity for certain configurations and numbers of electrons that make their orbits stable. Which chemical element an atom represents is determined by the number of protons in the nucleus; the neutral atom will have an equal number of electrons orbiting that nucleus. Individual chemical elements can create more stable electron configurations by combining to share their electrons. It is that sharing of electrons to create stable electronic orbits about the nucleus that appears to us as the chemistry of our macro world.

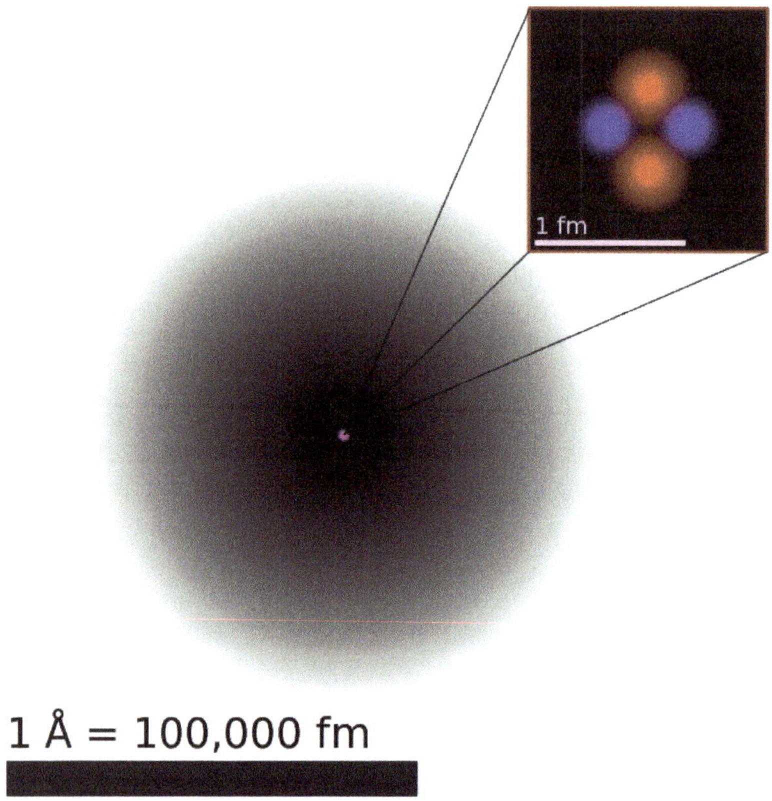

A figurative depiction of the helium-4 atom with the electron cloud in shades of gray. In the nucleus, the two protons and two neutrons are depicted in red and blue. This depiction shows the particles as separate, whereas in an actual helium atom, the protons are superimposed in space and most likely found at the very center of the nucleus, and the same is true of the two neutrons. Thus, all four particles are most likely found in exactly the same space, at the central point. Classical images of separate particles fail to model known charge distributions in very small nuclei. A more accurate image is that the spatial distribution of nucleons in a helium nucleus is much closer to the helium **electron cloud** shown here, although on a far smaller scale, than to the fanciful nucleus image.

Protons define the entire charge of a nucleus, and hence its chemical identity. Neutrons are electrically neutral, but contribute to the mass of a nucleus to nearly the same extent as the protons. Neutrons explain the phenomenon of isotopes – varieties of the same chemical element which differ only in their atomic mass, not their chemical action.

Composition and Shape

Protons and neutrons are fermions, with different values of the strong isospin quantum number, so two protons and two neutrons can share the same space wave function since they are not identical quantum entities. They are sometimes viewed as two different quantum states of the same particle, the nucleon. Two fermions, such as two protons, or two neutrons, or a proton + neutron (the deuteron) can exhibit bosonic behavior when they become loosely bound in pairs, which have integral spin.

In the rare case of a hypernucleus, a third baryon called a hyperon, containing one or more strange quarks and/or other unusual quark(s), can also share the wave function. However, this type of nucleus is extremely unstable and not found on Earth except in high energy physics experiments.

The neutron has a positively charged core of radius $\approx$ 0.3 fm surrounded by a compensating negative charge of radius between 0.3 fm and 2 fm. The proton has an approximately exponentially decaying positive charge distribution with a mean square radius of about 0.8 fm.

Nuclei can be spherical, rugby ball-shaped (prolate deformation), discus-shaped (oblate deformation), triaxial (a combination of oblate and prolate deformation) or pear-shaped.

Forces

Nuclei are bound together by the residual strong force (nuclear force). The residual strong force is a minor residuum of the strong interaction which binds quarks together to form protons and neutrons. This force is much weaker between neutrons and protons because it is mostly neutralized within them, in the same way that electromagnetic forces between neutral atoms (such as van der Waals forces that act between two inert gas atoms) are much weaker than the electromagnetic forces that hold the parts of the atoms together internally (for example, the forces that hold the electrons in an inert gas atom bound to its nucleus).

The nuclear force is highly attractive at the distance of typical nucleon separation, and this overwhelms the repulsion between protons due to the electromagnetic force, thus allowing nuclei to exist. However, the residual strong force has a limited range because it decays quickly with distance (see Yukawa potential); thus only nuclei smaller than a certain size can be completely stable. The largest known completely stable nucleus (i.e. stable to alpha, beta, and gamma decay) is lead-208 which contains a total of 208 nucleons (126 neutrons and 82 protons). Nuclei larger than this maximum are unstable and tend to be increasingly short-lived with larger numbers of nucleons. However, bismuth-209 is also stable to beta decay and has the longest half-life to alpha decay of any known isotope, estimated at a billion times longer than the age of the universe.

The residual strong force is effective over a very short range (usually only a few femtometres (fm); roughly one or two nucleon diameters) and causes an attraction between any pair of nucleons. For example, between protons and neutrons to form [NP] deuteron, and also between protons and protons, and neutrons and neutrons.

Halo Nuclei and Strong Force Range Limits

The effective absolute limit of the range of the strong force is represented by halo nuclei such as lithium-11 or boron-14, in which dineutrons, or other collections of neutrons, orbit at distances of about ten fermis (roughly similar to the 8 fermi radius of the nucleus of uranium-238). These nuclei are not maximally dense. Halo nuclei form at the extreme edges of the chart of the nuclides—the neutron drip line and proton drip line—and are all unstable with short half-lives, measured in milliseconds; for example, lithium-11 has a half-life of 8.8 milliseconds.

Halos in effect represent an excited state with nucleons in an outer quantum shell which has unfilled energy levels "below" it (both in terms of radius and energy). The halo may be made of either neutrons [NN, NNN] or protons [PP, PPP]. Nuclei which have a single neutron halo include ^{11}Be and ^{19}C. A two-neutron halo is exhibited by ^{6}He, ^{11}Li, ^{17}B, ^{19}B and ^{22}C. Two-neutron halo nuclei break into three fragments, never two, and are called Borromean nuclei because of this behavior (referring to a system of three interlocked rings in which breaking any ring frees both of the others). ^{8}He and ^{14}Be both exhibit a four-neutron halo. Nuclei which have a proton halo include ^{8}B and ^{26}P. A two-proton halo is exhibited by ^{17}Ne and ^{27}S. Proton halos are expected to be more rare and unstable than the neutron examples, because of the repulsive electromagnetic forces of the excess proton(s).

Nuclear Models

Although the standard model of physics is widely believed to completely describe the composition and behavior of the nucleus, generating predictions from theory is much more difficult than for most other areas of particle physics. This is due to two reasons:

- In principle, the physics within a nucleus can be derived entirely from quantum chromodynamics (QCD). In practice however, current computational and mathematical approaches for solving QCD in low-energy systems such as the nuclei are extremely limited. This is due to the phase transition that occurs between high-energy quark matter and low-energy hadronic matter, which renders perturbative techniques unusable, making it difficult to construct an accurate QCD-derived model of the forces between nucleons. Current approaches are limited to either phenomenological models such as the Argonne v18 potential or chiral effective field theory.
- Even if the nuclear force is well constrained, a significant amount of computational power is required to accurately compute the properties of nuclei ab initio. Developments in many-body theory have made this possible for many low mass and relatively stable nuclei, but further improvements in both computational power and mathematical approaches are required before heavy nuclei or highly unstable nuclei can be tackled.

Historically, experiments have been compared to relatively crude models that are necessarily imperfect. None of these models can completely explain experimental data on nuclear structure.

The nuclear radius (R) is considered to be one of the basic quantities that any model must predict. For stable nuclei (not halo nuclei or other unstable distorted nuclei) the nuclear radius is roughly proportional to the cube root of the mass number (A) of the nucleus, and particularly in nuclei containing many nucleons, as they arrange in more spherical configurations:

The stable nucleus has approximately a constant density and therefore the nuclear radius R can be approximated by the following formula,

$$R = r_0 A^{1/3}$$

where A = Atomic mass number (the number of protons Z, plus the number of neutrons N) and r_0 = 1.25 fm = 1.25×10^{-15} m. In this equation, the constant r_0 varies by 0.2 fm, depending on the nucleus in question, but this is less than 20% change from a constant.

In other words, packing protons and neutrons in the nucleus gives approximately the same total size result as packing hard spheres of a constant size (like marbles) into a tight spherical or almost spherical bag (some stable nuclei are not quite spherical, but are known to be prolate).

Models of nuclear structure include :

Liquid Drop Model

Early models of the nucleus viewed the nucleus as a rotating liquid drop. In this model, the trade-off of long-range electromagnetic forces and relatively short-range nuclear forces, together cause behavior which resembled surface tension forces in liquid drops of different sizes. This formula is successful at explaining many important phenomena of nuclei, such as their changing amounts of binding energy as their size and composition changes (see semi-empirical mass formula), but it does not explain the special stability which occurs when nuclei have special "magic numbers" of protons or neutrons.

The terms in the semi-empirical mass formula, which can be used to approximate the binding energy of many nuclei, are considered as the sum of five types of energies (see below). Then the picture of a nucleus as a drop of incompressible liquid roughly accounts for the observed variation of binding energy of the nucleus:

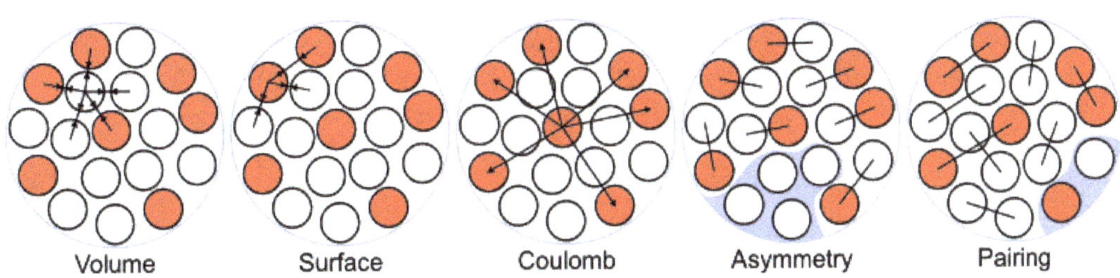

Volume energy. When an assembly of nucleons of the same size is packed together into the smallest volume, each interior nucleon has a certain number of other nucleons in contact with it. So, this nuclear energy is proportional to the volume.

Surface energy. A nucleon at the surface of a nucleus interacts with fewer other nucleons than one in the interior of the nucleus and hence its binding energy is less. This surface energy term takes that into account and is therefore negative and is proportional to the surface area.

Coulomb Energy. The electric repulsion between each pair of protons in a nucleus contributes toward decreasing its binding energy.

Asymmetry energy (also called Pauli Energy). An energy associated with the Pauli exclusion principle. Were it not for the Coulomb energy, the most stable form of nuclear matter would have the same number of neutrons as protons, since unequal numbers of neutrons and protons imply filling higher energy levels for one type of particle, while leaving lower energy levels vacant for the other type.

Pairing energy. An energy which is a correction term that arises from the tendency of proton pairs and neutron pairs to occur. An even number of particles is more stable than an odd number.

Shell Models and Other Quantum Models

A number of models for the nucleus have also been proposed in which nucleons occupy orbitals, much like the atomic orbitals in atomic physics theory. These wave models imagine nucleons to be either sizeless point particles in potential wells, or else probability waves as in the "optical model", frictionlessly orbiting at high speed in potential wells.

In the above models, the nucleons may occupy orbitals in pairs, due to being fermions, which allows to explain even/odd Z and N effects well-known from experiments. The exact nature and capacity of nuclear shells differs from those of electrons in atomic orbitals, primarily because the potential well in which the nucleons move (especially in larger nuclei) is quite different from the central electromagnetic potential well which binds electrons in atoms. Some resemblance to atomic orbital models may be seen in a small atomic nucleus like that of helium-4, in which the two protons and two neutrons separately occupy 1s orbitals analogous to the 1s orbital for the two electrons in the helium atom, and achieve unusual stability for the same reason. Nuclei with 5 nucleons are all extremely unstable and short-lived, yet, helium-3, with 3 nucleons, is very stable even with lack of a closed 1s orbital shell. Another nucleus with 3 nucleons, the triton hydrogen-3 is unstable and will decay into helium-3 when isolated. Weak nuclear stability with 2 nucleons {NP} in the 1s orbital is found in the deuteron hydrogen-2, with only one nucleon in each of the proton and neutron potential wells. While each nucleon is a fermion, the {NP} deuteron is a boson and thus does not follow Pauli Exclusion for close packing within shells. Lithium-6 with 6 nucleons is highly stable without a closed second 1p shell orbital. For light nuclei with total nucleon numbers 1 to 6 only those with 5 do not show some evidence of stability. Observations of beta-stability of light nuclei outside closed shells indicate that nuclear stability is much more complex than simple closure of shell orbitals with magic numbers of protons and neutrons.

For larger nuclei, the shells occupied by nucleons begin to differ significantly from electron shells, but nevertheless, present nuclear theory does predict the magic numbers of filled nuclear shells for both protons and neutrons. The closure of the stable shells predicts unusually stable configurations, analogous to the noble group of nearly-inert gases in chemistry. An example is the stability of the closed shell of 50 protons, which allows tin to have 10 stable isotopes, more than any other element. Similarly, the distance from shell-closure explains the unusual instability of isotopes which have far from stable numbers of these particles, such as the radioactive elements 43 (technetium) and 61 (promethium), each of which is preceded and followed by 17 or more stable elements.

There are however problems with the shell model when an attempt is made to account for nuclear properties well away from closed shells. This has led to complex post hoc distortions of the shape of the potential well to fit experimental data, but the question remains whether these mathematical manipulations actually correspond to the spatial deformations in real nuclei. Problems with the shell model have led some to propose realistic two-body and three-body nuclear force effects involving nucleon clusters and then build the nucleus on this basis. Two such cluster models are the Close-Packed Spheron Model of Linus Pauling and the 2D Ising Model of MacGregor.

Consistency between Models

As with the case of superfluid liquid helium, atomic nuclei are an example of a state in which both (1) “ordinary” particle physical rules for volume and (2) non-intuitive quantum mechanical rules for a wave-like nature apply. In superfluid helium, the helium atoms have volume, and essentially “touch” each other, yet at the same time exhibit strange bulk properties, consistent with a Bose–Einstein condensation. The latter reveals that they also have a wave-like nature and do not exhibit standard fluid properties, such as friction. For nuclei made of hadrons which are fermions, the same type of condensation does not occur, yet nevertheless, many nuclear properties can only be explained similarly by a combination of properties of particles with volume, in addition to the frictionless motion characteristic of the wave-like behavior of objects trapped in Erwin Schrödinger’s quantum orbitals.

EXOTIC ATOM

An exotic atom is an otherwise normal atom in which one or more sub-atomic particles have been replaced by other particles of the same charge. For example, electrons may be replaced by other negatively charged particles such as muons (muonic atoms) or pions (pionic atoms). Because these substitute particles are usually unstable, exotic atoms typically have very short lifetimes.

Positronium

Positronium (Ps) is a system consisting of an electron and its anti-particle, a positron, bound together into an exotic atom, specifically an onium. The system is unstable: the two particles annihilate each other to predominantly produce two or three gamma-rays, depending on the relative spin states. The orbit and energy levels of the two particles are similar to that of the hydrogen atom (which is a bound state of a proton and an electron). However, because of the reduced mass, the frequencies of the spectral lines are less than half of the corresponding hydrogen lines.

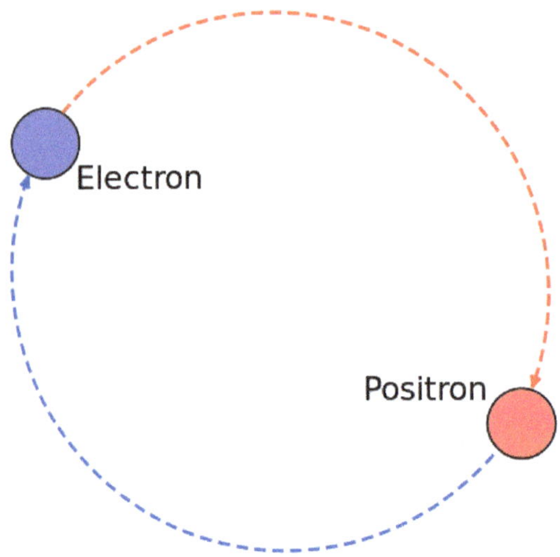

An electron and positron orbiting around their common centre of mass. This is a bound quantum state known as **positronium**.

States

The mass of positronium is 1.022 MeV, which is twice the electron mass minus the binding energy of a few eV. The ground state of positronium, like that of hydrogen, has two possible configurations depending on the relative orientations of the spins of the electron and the positron.

The singlet state, 1S0, with antiparallel spins ($S = 0$, $M_s = 0$) is known as para-positronium (p-Ps). It has a mean lifetime of 125 picoseconds and decays preferentially into two gamma rays with energy of 511 keV each (in the center-of-mass frame). By detecting these photons the location of the decay can be found. This process is used in positron-emission tomography. Para-positronium can decay into any even number of photons (2, 4, 6, ...), but the probability quickly decreases with the number: the branching ratio for decay into 4 photons is $1.439(2)\times10^{-6}$.

Para-positronium lifetime in vacuum is approximately

$$t_0 = \frac{2\hbar}{m_e c^2 \alpha^5} = 1.244 \times 10^{-10}\ \mathrm{s}.$$

The triplet state, $^3S_1$, with parallel spins ($S = 1$, $M_s = -1, 0, 1$) is known as ortho-positronium (o-Ps). It has a mean lifetime of 142.05 ± 0.02 ns, and the leading decay is three gammas. Other modes of decay are negligible; for instance, the five-photons mode has branching ratio of $\sim1.0\times10^{-6}$.

Ortho-positronium lifetime in vacuum can be calculated approximately as:

$$t_1 = \frac{\frac{1}{2}9h}{2m_e c^2 \alpha^6 (\pi^2 - 9)} = 1.386 \times 10^{-7} \text{ s.}$$

However more accurate calculations with corrections to order $O(\alpha^2)$ yield a value of 7.04 μs^{-1} for the decay rate, corresponding to a lifetime of 1.42×10^{-7} s.

Positronium in the 2S state is metastable having a lifetime of 1.1 μs against annihilation. The positronium created in such an excited state will quickly cascade down to the ground state, where annihilation will occur more quickly.

Measurements

Measurements of these lifetimes and energy levels have been used in precision tests of quantum electrodynamics, confirming quantum electrodynamics (QED) predictions to high precision. Annihilation can proceed via a number of channels, each producing gamma rays with total energy of 1022 keV (sum of the electron and positron mass-energy), usually 2 or 3, with up to 5 recorded.

The annihilation into a neutrino–antineutrino pair is also possible, but the probability is predicted to be negligible. The branching ratio for o-Ps decay for this channel is 6.2×10^{-18} (electron neutrino–antineutrino pair) and 9.5×10^{-21} (for other flavor) in predictions based on the Standard Model, but it can be increased by non-standard neutrino properties, like relatively high magnetic moment. The experimental upper limits on branching ratio for this decay (as well as for a decay into any "invisible" particles) are $<4.3\times10^{-7}$ for p-Ps and $<4.2\times10^{-7}$ for o-Ps.

Energy Levels

While precise calculation of positronium energy levels uses the Bethe–Salpeter equation or the Breit equation, the similarity between positronium and hydrogen allows a rough estimate. In this approximation, the energy levels are different because of a different effective mass, m*, in the energy equation (see electron energy levels for a derivation):

$$E_n = -\frac{\mu q_e^4}{8h^2 \varepsilon_0^2} \frac{1}{n^2},$$

where:

q_e is the charge magnitude of the electron (same as the positron),

h is Planck's constant,

ε_0 is the electric constant (otherwise known as the permittivity of free space),

μ is the reduced mass:

$$\mu = \frac{m_e m_p}{m_e + m_p} = \frac{m_e^2}{2m_e} = \frac{m_e}{2},$$

where m_e and m_p are, respectively, the mass of the electron and the positron (which are the same by definition as antiparticles).

Thus, for positronium, its reduced mass only differs from the electron by a factor of 2. This causes the energy levels to also roughly be half of what they are for the hydrogen atom.

So finally, the energy levels of positronium are given by

$$E_n = -\frac{1}{2}\frac{m_e q_e^4}{8h^2\varepsilon_0^2}\frac{1}{n^2} = \frac{-6.8\text{ eV}}{n^2}.$$

The lowest energy level of positronium (n = 1) is −6.8 electronvolts (eV). The next level is −1.7 eV. The negative sign implies a bound state. Positronium can also be considered by a particular form of the two-body Dirac equation; Two point particles with a Coulomb interaction can be exactly separated in the (relativistic) center-of-momentum frame and the resulting ground-state energy has been obtained very accurately using finite element methods of J. Shertzer. The Dirac equation whose Hamiltonian comprises two Dirac particles and a static Coulomb potential is not relativistically invariant. But if one adds the $1/c^{2n}$ (or α^{2n}, where α is the fine-structure constant) terms, where n = 1,2…, then the result is relativistically invariant. Only the leading term is included. The α^2 contribution is the Breit term; workers rarely go to α^4 because at α^3 one has the Lamb shift, which requires quantum electrodynamics.

History

Croatian scientist Stjepan Mohorovičić predicted the existence of positronium in a 1934 article published in Astronomische Nachrichten, in which he called it the "electrum". Other sources credit Carl Anderson as having predicted its existence in 1932 while at Caltech. It was experimentally discovered by Martin Deutsch at MIT in 1951 and became known as positronium. Many subsequent experiments have precisely measured its properties and verified predictions of QED.

The Positronium Beam at University College London, a lab used to study the properties of positronium

There was a discrepancy known as the ortho-positronium lifetime puzzle that persisted for some time, but was eventually resolved with further calculations and measurements. Measurements were in error because of the lifetime measurement of unthermalised positronium, which was only produced at a small rate. This had yielded lifetimes that were too long. Also calculations using relativistic QED are difficult to perform, so they had been done to only the first order. Corrections that involved higher orders were then calculated in a non-relativistic QED.

Exotic Compounds

Molecular bonding was predicted for positronium. Molecules of positronium hydride (PsH) can be made. Positronium can also form a cyanide and can form bonds with halogens or lithium.

The first observation of di-positronium molecules—molecules consisting of two positronium atoms—was reported on 12 September 2007 by David Cassidy and Allen Mills from University of California, Riverside.

Natural Occurrence

Positronium in high energy states has been predicted to be the dominant form of atomic matter in the universe in the far future, if proton decay is a reality.

Muonium

Muonium is an exotic atom made up of an antimuon and an electron, which was discovered in 1960 and is given the chemical symbol Mu. During the muon's 2.2 μs lifetime, muonium can enter into compounds such as muonium chloride (MuCl) or sodium muonide (NaMu). Due to the mass difference between the antimuon and the electron, muonium (μ+e−) is more similar to atomic hydrogen (p+e−) than positronium (e+e−). Its Bohr radius and ionization energy are within 0.5% of hydrogen, deuterium, and tritium, and thus it can usefully be considered as an exotic light isotope of hydrogen.

Although muonium is short-lived, physical chemists study it using muon spin spectroscopy (μSR), a magnetic resonance technique analogous to nuclear magnetic resonance (NMR) or electron spin resonance (ESR) spectroscopy. Like ESR, μSR is useful for the analysis of chemical transformations and the structure of compounds with novel or potentially valuable electronic properties. Muonium is usually studied by muon spin rotation, in which the Mu atom's spin precesses in a magnetic field applied transverse to the muon spin direction (since muons are typically produced in a spin-polarized state from the decay of pions), and by avoided level crossing (ALC), which is also called level crossing resonance (LCR). The latter employs a magnetic field applied longitudinally to the polarization direction, and monitors the relaxation of muon spins caused by "flip/flop" transitions with other magnetic nuclei.

Because the muon is a lepton, the atomic energy levels of muonium can be calculated with great precision from quantum electrodynamics (QED), unlike in the case of hydrogen, where the precision is limited by uncertainties related to the internal structure of the proton. For this reason, muonium is an ideal system for studying bound-state QED and also for searching for physics beyond the standard model.

Nomenclature

Normally in the nomenclature of particle physics, an atom composed of a positively charged particle bound to an electron is named after the positive particle with "-ium" appended, in this case "muium". The suffix "-onium" is mostly used for bound states of a particle with its own antiparticle. The exotic atom consisting of a muon and an antimuon is known as "true muonium". It is yet to be observed, but it may have been generated in the collision of electron and positron beams.

Tau (Particle)

The **tau** (τ), also called the tau lepton, tau particle, or tauon, is an elementary particle similar to the electron, with negative electric charge and a spin of 1/2. Together with the electron, the muon, and the three neutrinos, it is a lepton. Like all elementary particles with half-integral spin, the tau has a corresponding antiparticle of opposite charge but equal mass and spin, which in the tau's case is the antitau (also called the positive tau). Tau particles are denoted by τ− and the antitau by τ+.

Tau leptons have a lifetime of 2.9×10^{-13} s and a mass of 1776.82 MeV/c^2 (compared to 105.7 MeV/c^2 for muons and 0.511 MeV/c^2 for electrons). Since their interactions are very similar to those of the electron, a tau can be thought of as a much heavier version of the electron. Because of their greater mass, tau particles do not emit as much bremsstrahlung radiation as electrons; consequently they are potentially highly penetrating, much more so than electrons. However, because of their short lifetime, the range of the tau is mainly set by their decay length, which is too small for bremsstrahlung to be noticeable: their penetrating power appears only at ultra high energy (above PeV energies).

As with the case of the other charged leptons, the tau has an associated tau neutrino, denoted by ντ.

History

The tau was detected in a series of experiments between 1974 and 1977 by Martin Lewis Perl with his colleagues at the SLAC-LBL group. Their equipment consisted of SLAC's then-new e+– e− colliding ring, called SPEAR, and the LBL magnetic detector. They could detect and distinguish between leptons, hadrons and photons. They did not detect the tau directly, but rather discovered anomalous events:

We have discovered 64 events of the form $e^+ + e^- \rightarrow e^\pm + \mu^\mp +$ at least two undetected particles for which we have no conventional explanation.

The need for at least two undetected particles was shown by the inability to conserve energy and momentum with only one. However, no other muons, electrons, photons, or hadrons were detected. It was proposed that this event was the production and subsequent decay of a new particle pair:

$$e^+ + e^- \rightarrow \tau^+ + \tau^- \rightarrow e^\pm + \mu^\mp + 4\nu$$

This was difficult to verify, because the energy to produce the τ+τ− pair is similar to the threshold for D meson production. Work done at DESY-Hamburg, and with the Direct Electron Counter (DELCO) at SPEAR, subsequently established the mass and spin of the tau.

The symbol τ was derived from the Greek τρίτον (triton, meaning "third" in English), since it was the third charged lepton discovered.

Martin Perl shared the 1995 Nobel Prize in Physics with Frederick Reines. The latter was awarded his share of the prize for experimental discovery of the neutrino.

Tau Decay

The tau is the only lepton that can decay into hadrons – the other leptons do not have the necessary mass. Like the other decay modes of the tau, the hadronic decay is through the weak interaction.

The branching ratio of the dominant hadronic tau decays are:

- 25.52% for decay into a charged pion, a neutral pion, and a tau neutrino;
- 10.83% for decay into a charged pion and a tau neutrino;
- 9.30% for decay into a charged pion, two neutral pions, and a tau neutrino;
- 8.99% for decay into three charged pions (of which two have the same electrical charge) and a tau neutrino;
- 2.70% for decay into three charged pions (of which two have the same electrical charge), a neutral pion, and a tau neutrino;
- 1.05% for decay into three neutral pions, a charged pion, and a tau neutrino.

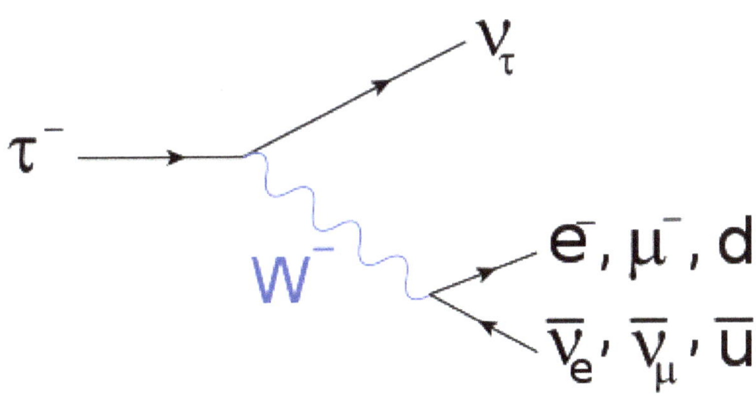

Feynman diagram of the common decays of the tau by emission of a W boson.

In total, the tau lepton will decay hadronically approximately 64.79% of the time.

Since the tauonic lepton number is conserved in weak decays, a tau neutrino is always created when a tau decays.

The branching ratio of the common purely leptonic tau decays are:

- 17.82% for decay into a tau neutrino, electron and electron antineutrino;
- 17.39% for decay into a tau neutrino, muon and muon antineutrino.

The similarity of values of the two branching ratios is a consequence of lepton universality.

The tau lepton is predicted to form exotic atoms like other charged subatomic particles. One of such, called **tauonium** by the analogy to muonium, consists in antitauon and an electron: τ+e−.

Another one is an onium atom τ+τ− called true tauonium and is difficult to detect due to tau's extremely short lifetime at low (non-relativistic) energies needed to form this atom. Its detection is important for quantum electrodynamics.

Onium

An onium (plural: onia) is a bound state of a particle and its antiparticle. They are usually named by adding the suffix -onium to the name of the constituting particle except for muonium which, despite its name, not a bound muon–antimuon onium, but an electron–antimuon bound state, and whose name was assigned by IUPAC. A muon–antimuon onium would be named true muonium or muononium.

Examples

Positronium is an onium which consists of an electron and a positron bound together as a long-lived metastable state. Positronium has been studied since the 1950s to understand bound states in quantum field theory. A recent development called non-relativistic quantum electrodynamics (NRQED) used this system as a proving ground.

Pionium, a bound state of two oppositely-charged pions, is interesting for exploring the strong interaction. This should also be true of protonium. The true analogs of positronium in the theory of strong interactions are the quarkonium states: they are mesons made of a heavy quark and antiquark (namely, charmonium and bottomonium). Exploration of these states through non-relativistic quantum chromodynamics (NRQCD) and lattice QCD are increasingly important tests of quantum chromodynamics.

Understanding bound states of hadrons such as pionium and protonium is also important in order to clarify notions related to exotic hadrons such as mesonic molecules and pentaquark states.

SUPERATOM

A superatom is any cluster of atoms that seem to exhibit some of the properties of elemental atoms.

Sodium atoms, when cooled from vapor, naturally condense into clusters, preferentially containing a magic number of atoms (2, 8, 20, 40, 58, etc.). The first two of these can be recognized as the numbers of electrons needed to fill the first and second shells, respectively. The superatom suggestion is that free electrons in the cluster occupy a new set of orbitals that are defined by the entire group of atoms, i.e. cluster, rather than each individual atom separately (non-spherical or doped clusters show deviations in the number of electrons that form a closed shell as the potential is defined by the shape of the positive nuclei.) Superatoms tend to behave chemically in a way that will allow them to have a closed shell of electrons, in this new counting scheme. Therefore, a superatom with one more electron than a full shell should give up that electron very easily, similar to an alkali metal, and a cluster with one electron short of full shell should have a large electron affinity, such as a halogen.

Aluminium Clusters

Certain aluminium clusters have superatom properties. These aluminium clusters are generated as anions (Al_n^- with n = 1,2,3, ...) in helium gas and reacted with a gas containing iodine. When analyzed by mass spectrometry one main reaction product turns out to be $Al_{13}I^-$. These clusters of 13 aluminium atoms with an extra electron added do not appear to react with oxygen when it is introduced in the same gas stream. Assuming each atom liberates its 3 valence electrons, this means that there are 40 electrons present, which is one of the magic numbers noted above for sodium, and implies that these numbers are a reflection of the noble gases. Calculations show that the additional electron is located in the aluminium cluster at the location directly opposite from the iodine atom. The cluster must therefore have a higher electron affinity for the electron than iodine and therefore the aluminium cluster is called a superhalogen. The cluster component in $Al_{13}I^-$ ion is similar to an iodide ion or better still a bromide ion. The related $Al_{13}I_2^-$ cluster is expected to behave chemically like the triiodide ion.

Similarly it has been noted that Al_{14} clusters with 42 electrons (2 more than the magic numbers) appear to exhibit the properties of an alkaline earth metal which typically adopt +2 valence states. This is only known to occur when there are at least 3 iodine atoms attached to an Al_{14}^- cluster, $Al_{14}I_3^-$. The anionic cluster has a total of 43 itinerant electrons, but the three iodine atoms each remove one of the itinerant electrons to leave 40 electrons in the jellium shell.

It is particularly easy and reliable to study atomic clusters of inert gas atoms by computer simulation because interaction between two atoms can be approximated very well by the Lennard-Jones potential. Other methods are readily available and it has been established that the magic numbers are 13, 19, 23, 26, 29, 32, 34, 43, 46, 49, 55, etc.

- Al_7 = the property is similar to germanium atoms.
- Al_{13} = the property is similar to halogen atoms, more specifically, chlorine.
- $Al_{13}I_x^-$, where x = 1–13.
- Al_{14} = the property is similar to alkaline earth metals.
- $Al_{14}I_x^-$, where x = 1–14.
- Al_{23}

- Al_{37}

Other Clusters

- $Li(HF)_3Li$ = the $(HF)_3$ interior causes 2 valence electrons from the Li to orbit the entire molecule as if it were an atom's nucleus.
- $VSi_{16}F$ = has ionic bonding.
- A cluster of 13 platinum becomes paramagnetic.
- A cluster of 2000 rubidium atoms.

Superatom Complexes

Superatom complexes are a special group of superatoms that incorporate a metal core which is stabilized by organic ligands. In thiolate-protected gold cluster complexes a simple electron counting rule can be used to determine the total number of electrons (n_e) which correspond to a magic number via,

$$n_e = N\nu_A - M - z$$

where N is the number of metal atoms (A) in the core, v is the atomic valence, M is the number of electron withdrawing ligands, and z is the overall charge on the complex. For example the $Au_{102}(p\text{-}MBA)_{44}$ has 58 electrons and corresponds to a closed shell magic number.

Gold superatom complexes

- $Au_{25}(SMe)_{18}^{-}$
- $Au_{102}(p\text{-}MBA)_{44}$
- $Au_{144}(SR)_{60}$

Other superatom complexes

- $Ga_{23}(N(Si(CH_3)_3)_2)_{11}$
- $Al_{50}(C_5(CH_3)_5)_{12}$

MOLECULE

A molecule is an electrically neutral group of two or more atoms held together by chemical bonds. Molecules are distinguished from ions by their lack of electrical charge. However, in quantum physics, organic chemistry, and biochemistry, the term molecule is often used less strictly, also being applied to polyatomic ions.

In the kinetic theory of gases, the term molecule is often used for any gaseous particle regardless of its composition. According to this definition, noble gas atoms are considered molecules as they are in fact monoatomic molecules.

A molecule may be homonuclear, that is, it consists of atoms of one chemical element, as with oxygen (O_2); or it may be heteronuclear, a chemical compound composed of more than one element, as with water (H_2O). Atoms and complexes connected by non-covalent interactions, such as hydrogen bonds or ionic bonds, are generally not considered single molecules.

Molecules as components of matter are common in organic substances (and therefore biochemistry). They also make up most of the oceans and atmosphere. However, the majority of familiar solid substances on Earth, including most of the minerals that make up the crust, mantle, and core of the Earth, contain many chemical bonds, but are not made of identifiable molecules. Also, no typical molecule can be defined for ionic crystals (salts) and covalent crystals (network solids), although these are often composed of repeating unit cells that extend either in a plane (such as in graphene) or three-dimensionally (such as in diamond, quartz, or sodium chloride). The theme of repeated unit-cellular-structure also holds for most condensed phases with metallic bonding, which means that solid metals are also not made of molecules. In glasses (solids that exist in a vitreous disordered state), atoms may also be held together by chemical bonds with no presence of any definable molecule, nor any of the regularity of repeating units that characterizes crystals.

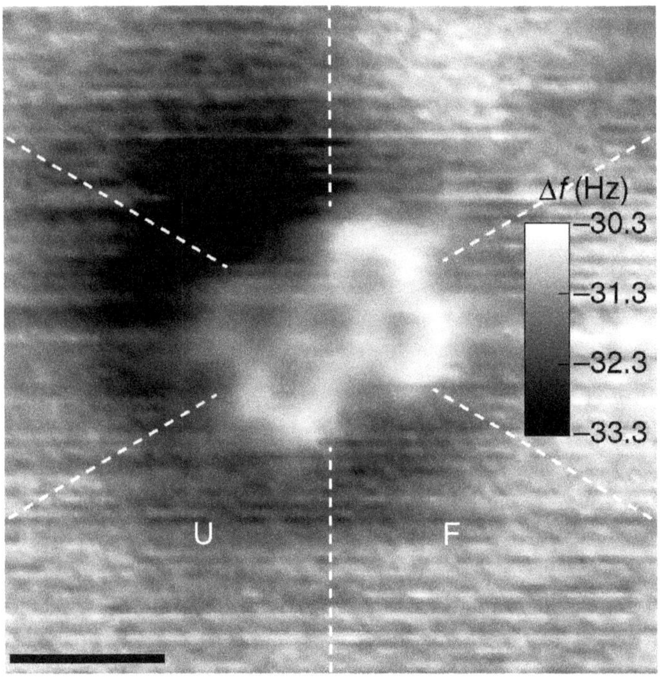

Atomic force microscopy image of a PTCDA molecule, which contains five carbon rings in a non-linear arrangement.

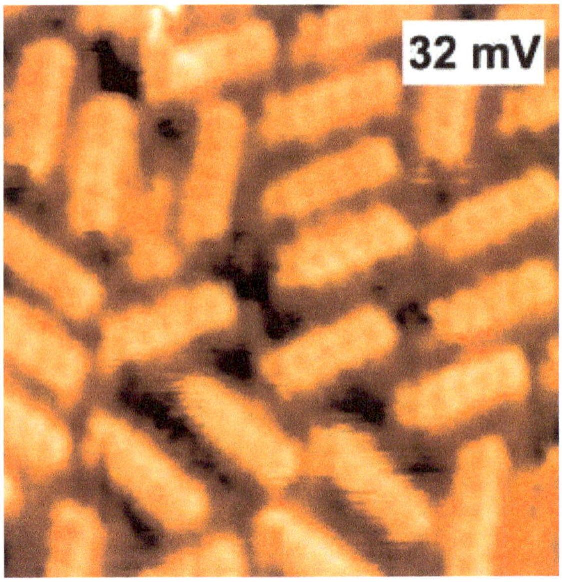

A scanning tunneling microscopy image of pentacene molecules, which consist of linear chains of five carbon rings.

Molecular Science

The science of molecules is called molecular chemistry or molecular physics, depending on whether the focus is on chemistry or physics. Molecular chemistry deals with the laws governing the interaction between molecules that results in the formation and breakage of chemical bonds, while molecular physics deals with the laws governing their structure and properties. In practice, however, this distinction is vague. In molecular sciences, a molecule consists of a stable system (bound state) composed of two or more atoms. Polyatomic ions may sometimes be usefully thought of as electrically charged molecules. The term unstable molecule is used for very reactive species, i.e., short-lived assemblies (resonances) of electrons and nuclei, such as radicals, molecular ions, Rydberg molecules, transition states, van der Waals complexes, or systems of colliding atoms as in Bose–Einstein condensate.

History and Etymology

According to Merriam-Webster and the Online Etymology Dictionary, the word "molecule" derives from the Latin "moles" or small unit of mass.

- Molecule (1794) – "extremely minute particle", from French molécule (1678), from New Latin molecula, diminutive of Latin moles "mass, barrier". A vague meaning at first; the vogue for the word (used until the late 18th century only in Latin form) can be traced to the philosophy of Descartes.

The definition of the molecule has evolved as knowledge of the structure of molecules has increased. Earlier definitions were less precise, defining molecules as the smallest particles of pure chemical substances that still retain their composition and chemical properties. This definition often breaks down since many substances in ordinary experience, such as rocks, salts, and metals, are composed of large crystalline networks of chemically bonded atoms or ions, but are not made of discrete molecules.

Bonding

Molecules are held together by either covalent bonding or ionic bonding. Several types of non-metal elements exist only as molecules in the environment. For example, hydrogen only exists as hydrogen molecule. A molecule of a compound is made out of two or more elements.

Covalent

A covalent bond is a chemical bond that involves the sharing of electron pairs between atoms. These electron pairs are termed shared pairs or bonding pairs, and the stable balance of attractive and repulsive forces between atoms, when they share electrons, is termed covalent bonding.

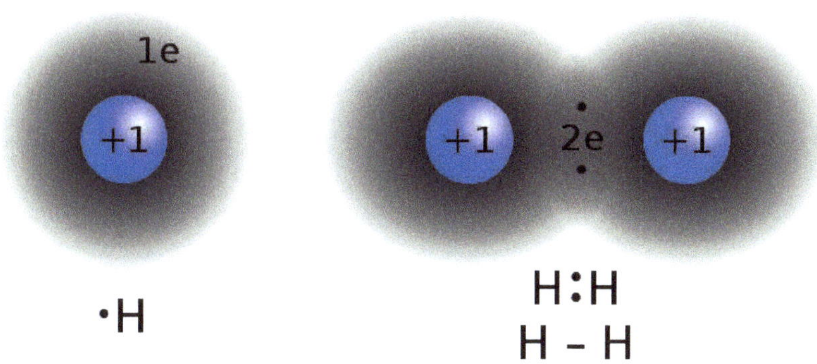

A covalent bond forming H_2 (right) where two hydrogen atoms share the two electrons

Ionic

Ionic bonding is a type of chemical bond that involves the electrostatic attraction between oppositely charged ions, and is the primary interaction occurring in ionic compounds. The ions are atoms that have lost one or more electrons (termed cations) and atoms that have gained one or more electrons (termed anions). This transfer of electrons is termed electrovalence in contrast to covalence. In the simplest case, the cation is a metal atom and the anion is a nonmetal atom, but these ions can be of a more complex nature, e.g. molecular ions like NH_4^+ or SO_4^{2-}. In simpler words, an ionic bond is the transfer of electrons from a metal to a non-metal for both atoms to obtain a full valence shell.

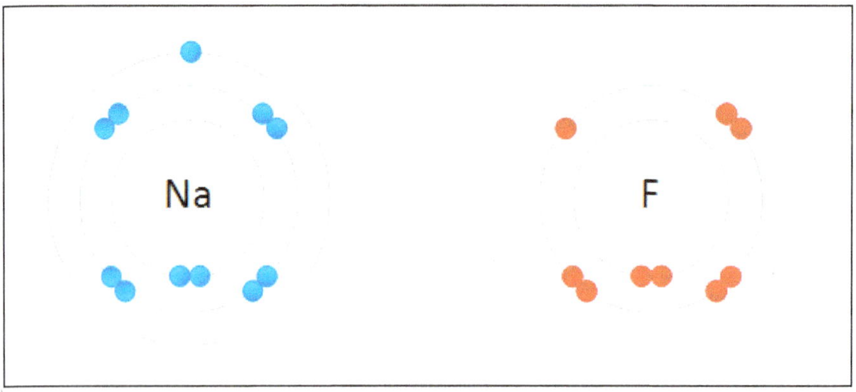

Sodium and fluorine undergoing a redox reaction to form sodium fluoride. Sodium loses its outer electron to give it a stable electron configuration, and this electron enters the fluorine atom exothermically.

Molecular Size

Most molecules are far too small to be seen with the naked eye, but there are exceptions. DNA, a macromolecule, can reach macroscopic sizes, as can molecules of many polymers. Molecules commonly used as building blocks for organic synthesis have a dimension of a few angstroms (Å) to several dozen Å, or around one billionth of a meter. Single molecules cannot usually be observed by light (as noted above), but small molecules and even the outlines of individual atoms may be traced in some circumstances by use of an atomic force microscope. Some of the largest molecules are macromolecules or supermolecules.

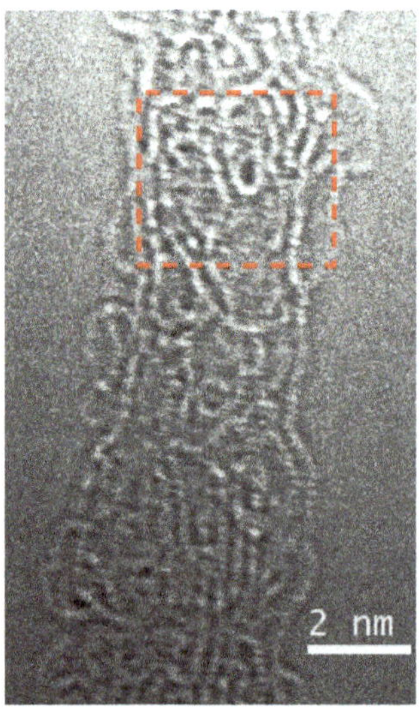

Arrangement of polyvinylidene fluoride molecules in a nanofiber – transmission electron microscopy image.

The smallest molecule is the diatomic hydrogen (H_2), with a bond length of 0.74 Å.

Effective molecular radius is the size a molecule displays in solution. The table of permselectivity for different substances contains examples.

Molecular Formulas

Chemical Formula Types

The chemical formula for a molecule uses one line of chemical element symbols, numbers, and sometimes also other symbols, such as parentheses, dashes, brackets, and plus (+) and minus (−) signs. These are limited to one typographic line of symbols, which may include subscripts and superscripts.

A compound's empirical formula is a very simple type of chemical formula. It is the simplest integer ratio of the chemical elements that constitute it. For example, water is always composed of a 2:1 ratio of hydrogen to oxygen atoms, and ethyl alcohol or ethanol is always composed of carbon, hydrogen, and oxygen in a 2:6:1 ratio. However, this does not determine the kind of molecule uniquely – dimethyl ether has the same ratios as ethanol, for instance. Molecules with the same atoms in different arrangements are called isomers. Also carbohydrates, for example, have the same ratio (carbon:hydrogen:oxygen= 1:2:1) (and thus the same empirical formula) but different total numbers of atoms in the molecule.

The molecular formula reflects the exact number of atoms that compose the molecule and so characterizes different molecules. However different isomers can have the same atomic composition while being different molecules.

The empirical formula is often the same as the molecular formula but not always. For example, the molecule acetylene has molecular formula C_2H_2, but the simplest integer ratio of elements is CH.

The molecular mass can be calculated from the chemical formula and is expressed in conventional atomic mass units equal to 1/12 of the mass of a neutral carbon-12 (^{12}C isotope) atom. For network solids, the term formula unit is used in stoichiometric calculations.

Structural Formula

For molecules with a complex 3-dimensional structure, especially involving atoms bonded to four different substituents, a simple molecular formula or even semi-structural chemical formula may not be enough to completely specify the molecule. In this case, a graphical type of formula called a structural formula may be needed. Structural formulas may in turn be represented with a one-dimensional chemical name, but such chemical nomenclature requires many words and terms which are not part of chemical formulas.

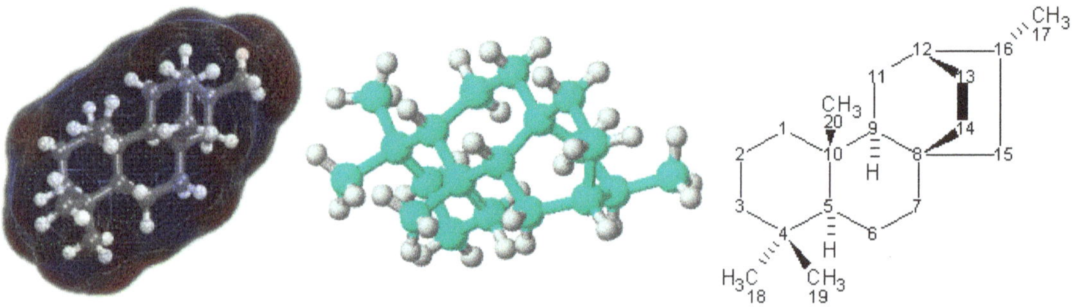

3D (left and center) and 2D (right) representations of the terpenoid molecule atisane

Molecular Geometry

Molecules have fixed equilibrium geometries—bond lengths and angles— about which they continuously oscillate through vibrational and rotational motions. A pure substance is composed of molecules with the same average geometrical structure.

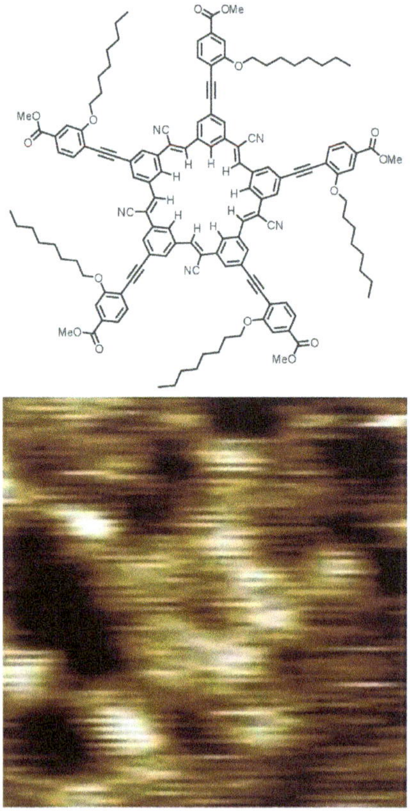

Structure and STM image of a "cyanostar" dendrimer molecule.

The chemical formula and the structure of a molecule are the two important factors that determine its properties, particularly its reactivity. Isomers share a chemical formula but normally have very different properties because of their different structures. Stereoisomers, a particular type of isomer, may have very similar physico-chemical properties and at the same time different biochemical activities.

Molecular Spectroscopy

Molecular spectroscopy deals with the response (spectrum) of molecules interacting with probing signals of known energy (or frequency, according to Planck's formula). Molecules have quantized energy levels that can be analyzed by detecting the molecule's energy exchange through absorbance or emission. Spectroscopy does not generally refer to diffraction studies where particles such as neutrons, electrons, or high energy X-rays interact with a regular arrangement of molecules (as in a crystal).

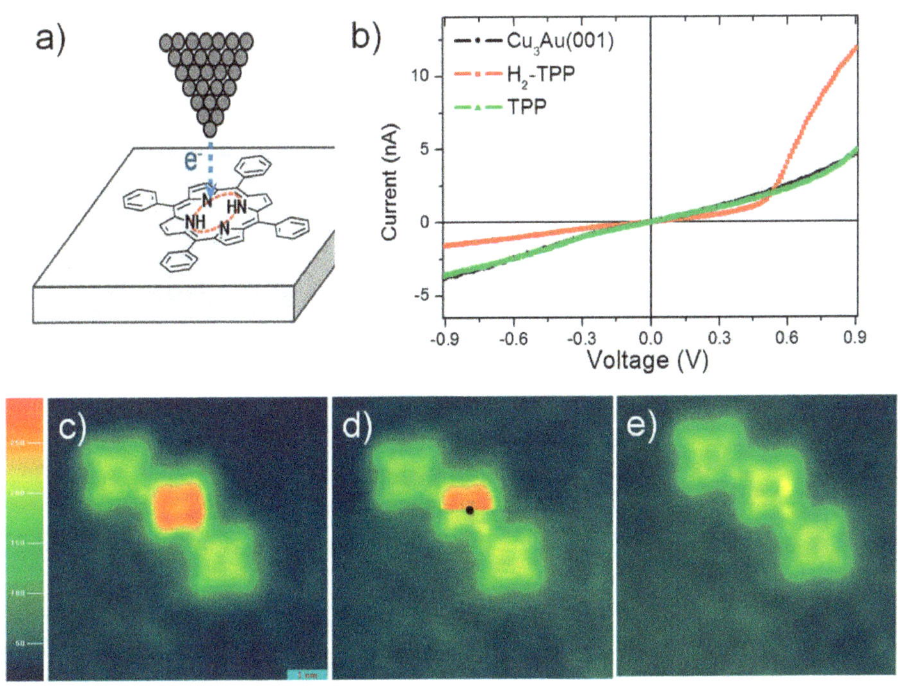

Hydrogen can be removed from individual H_2TPP molecules by applying excess voltage to the tip of a scanning tunneling microscope (STM, a); this removal alters the current-voltage (I-V) curves of TPP molecules, measured using the same STM tip, from diode like (red curve in b) to resistor like (green curve). Image (c) shows a row of TPP, H_2TPP and TPP molecules. While scanning image (d), excess voltage was applied to H_2TPP at the black dot, which instantly removed hydrogen, as shown in the bottom part of (d) and in the rescan image (e). Such manipulations can be used in single-molecule electronics.

Microwave spectroscopy commonly measures changes in the rotation of molecules, and can be used to identify molecules in outer space. Infrared spectroscopy measures changes in vibration of molecules, including stretching, bending or twisting motions. It is commonly used to identify the kinds of bonds or functional groups in molecules. Changes in the arrangements of electrons yield absorption or emission lines in ultraviolet, visible or near infrared light, and result in colour. Nuclear resonance spectroscopy actually measures the environment of particular nuclei in the molecule, and can be used to characterise the numbers of atoms in different positions in a molecule.

Theoretical Aspects

The study of molecules by molecular physics and theoretical chemistry is largely based on quantum mechanics and is essential for the understanding of the chemical bond. The simplest of molecules is the hydrogen molecule-ion, H_2^+, and the simplest of all the chemical bonds is the one-electron bond. H_2^+ is composed of two positively charged protons and one negatively charged electron, which means that the Schrödinger equation for the system can be solved more easily due to the lack of electron–electron repulsion. With the development of fast digital computers, approximate solutions for more complex molecules became possible and are one of the main aspects of computational chemistry.

When trying to define rigorously whether an arrangement of atoms is sufficiently stable to be considered a molecule, IUPAC suggests that it "must correspond to a depression on the potential energy surface that is deep enough to confine at least one vibrational state". This definition does not depend on the nature of the interaction between the atoms, but only on the strength of the interaction. In fact, it includes weakly bound species that would not traditionally be considered molecules, such as the helium dimer, He_2, which has one vibrational bound state and is so loosely bound that it is only likely to be observed at very low temperatures.

Whether or not an arrangement of atoms is sufficiently stable to be considered a molecule is inherently an operational definition. Philosophically, therefore, a molecule is not a fundamental entity (in contrast, for instance, to an elementary particle); rather, the concept of a molecule is the chemist's way of making a useful statement about the strengths of atomic-scale interactions in the world that we observe.

REFERENCES

1. "Atom". Compendium of Chemical Terminology (IUPAC Gold Book) (2nd ed.). IUPAC. Retrieved 2015-04-25.
2. Ghosh, D. C.; Biswas, R. (2002). "Theoretical calculation of Absolute Radii of Atoms and Ions. Part 1. The Atomic Radii". Int. J. Mol. Sci. 3: 87–113. doi:10.3390/i3020087Freely accessible.
3. Leigh, G. J., ed. (1990). International Union of Pure and Applied Chemistry, Commission on the Nomenclature of Inorganic Chemistry, Nomenclature of Organic Chemistry – Recommendations 1990. Oxford: Blackwell Scientific Publications. p. 35. ISBN 0-08-022369-9. An atom is the smallest unit quantity of an element that is capable of existence whether alone or in chemical combination with other atoms of the same or other elements.

4. Andrew G. van Melsen (1952). From Atomos to Atom. Mineola, N.Y.: Dover Publications. ISBN 0-486-49584-1.
5. Dalton, John. "On the Absorption of Gases by Water and Other Liquids", in Memoirs of the Literary and Philosophical Society of Manchester. 1803. Retrieved on August 29, 2007.
6. Einstein, Albert (1905). "Über die von der molekularkinetischen Theorie der Wärme geforderte Bewegung von in ruhenden Flüssigkeiten suspendierten Teilchen" (PDF). Annalen der Physik (in German). 322 (8): 549–560. Bibcode:1905AnP...322..549E. doi:10.1002/andp.19053220806. Retrieved 4 February 2007.
7. Mazo, Robert M. (2002). Brownian Motion: Fluctuations, Dynamics, and Applications. Oxford University Press. pp. 1–7. ISBN 0-19-851567-7. OCLC 48753074.
8. Lee, Y.K.; Hoon, K. (1995). "Brownian Motion". Imperial College. Archived from the original on 18 December 2007. Retrieved 18 December 2007.
9. Patterson, G. (2007). "Jean Perrin and the triumph of the atomic doctrine". Endeavour. 31 (2): 50–53. doi:10.1016/j.endeavour.2007.05.003. PMID 17602746.
10. Thomson, J. J. (August 1901). "On bodies smaller than atoms". The Popular Science Monthly. Bonnier Corp.: 323–335. Retrieved 2009-06-21.
11. "J.J. Thomson". Nobel Foundation. 1906. Retrieved 20 December 2007.
12. Rutherford, E. (1911). "The Scattering of α and β Particles by Matter and the Structure of the Atom" (PDF). Philosophical Magazine. 21 (125): 669–88. doi:10.1080/14786440508637080.
13. "The Gold Foil Experiment". myweb.usf.edu.
14. "Frederick Soddy, The Nobel Prize in Chemistry 1921". Nobel Foundation. Retrieved 18 January 2008.
15. Thomson, Joseph John (1913). "Rays of positive electricity". Proceedings of the Royal Society. A. 89 (607): 1–20. Bibcode:1913RSPSA..89....1T. doi:10.1098/rspa.1913.0057Freely accessible.
16. Stern, David P. (16 May 2005). "The Atomic Nucleus and Bohr's Early Model of the Atom". NASA/Goddard Space Flight Center. Retrieved 20 December 2007.
17. Bohr, Niels (11 December 1922). "Niels Bohr, The Nobel Prize in Physics 1922, Nobel Lecture". Nobel Foundation. Retrieved 16 February 2008.
18. Pais, Abraham (1986). Inward Bound: Of Matter and Forces in the Physical World. New York: Oxford University Press. pp. 228–230. ISBN 0-19-851971-0.
19. Lewis, Gilbert N. (1916). "The Atom and the Molecule". Journal of the American Chemical Society. 38 (4): 762–786. doi:10.1021/ja02261a002.
20. Scerri, Eric R. (2007). The periodic table: its story and its significance. Oxford University Press US. pp. 205–226. ISBN 0-19-530573-6.
21. Langmuir, Irving (1919). "The Arrangement of Electrons in Atoms and Molecules". Journal of the American Chemical Society. 41 (6): 868–934. doi:10.1021/

ja02227a002.

22. Scully, Marlan O.; Lamb, Willis E.; Barut, Asim (1987). "On the theory of the Stern-Gerlach apparatus". Foundations of Physics. 17 (6): 575–583. Bibcode:1987FoPh...17..575S. doi:10.1007/BF01882788.
23. What is the Heisenberg Uncertainty Principle? - Chad Orzel TED-ED talk
24. Brown, Kevin (2007). "The Hydrogen Atom". MathPages. Retrieved 21 December 2007.
25. Harrison, David M. (2000). "The Development of Quantum Mechanics". University of Toronto. Archived from the original on 25 December 2007. Retrieved 21 December 2007.
26. Aston, Francis W. (1920). "The constitution of atmospheric neon". Philosophical Magazine. 39 (6): 449–55. doi:10.1080/14786440408636058.
27. Chadwick, James (12 December 1935). "Nobel Lecture: The Neutron and Its Properties". Nobel Foundation. Retrieved 21 December 2007.
28. Bowden, Mary Ellen (1997). "Otto Hahn, Lise Meitner, and Fritz Strassmann". Chemical achievers : the human face of the chemical sciences. Philadelphia, PA: Chemical Heritage Foundation. pp. 76–80, 125. ISBN 9780941901123.
29. "Otto Hahn, Lise Meitner, and Fritz Strassmann". Chemical Heritage Foundation. Retrieved 27 October 2016.
30. Meitner, Lise; Frisch, Otto Robert (1939). "Disintegration of uranium by neutrons: a new type of nuclear reaction". Nature. 143 (3615): 239–240. Bibcode:1939Natur.143..239M. doi:10.1038/143239a0.
31. Schroeder, M. "Lise Meitner – Zur 125. Wiederkehr Ihres Geburtstages" (in German). Retrieved 4 June 2009.
32. Crawford, E.; Sime, Ruth Lewin; Walker, Mark (1997). "A Nobel tale of postwar injustice". Physics Today. 50 (9): 26–32. Bibcode:1997PhT....50i..26C. doi:10.1063/1.881933.
33. Kullander, Sven (28 August 2001). "Accelerators and Nobel Laureates". Nobel Foundation. Retrieved 31 January 2008.
34. "The Nobel Prize in Physics 1990". Nobel Foundation. 17 October 1990. Retrieved 31 January 2008.
35. Demtröder, Wolfgang (2002). Atoms, Molecules and Photons: An Introduction to Atomic- Molecular- and Quantum Physics (1st ed.). Springer. pp. 39–42. ISBN 3-540-20631-0. OCLC 181435713.
36. Woan, Graham (2000). The Cambridge Handbook of Physics. Cambridge University Press. p. 8. ISBN 0-521-57507-9. OCLC 224032426.
37. MacGregor, Malcolm H. (1992). The Enigmatic Electron. Oxford University Press. pp. 33–37. ISBN 0-19-521833-7. OCLC 223372888.
38. Particle Data Group (2002). "The Particle Adventure". Lawrence Berkeley Laboratory. Archived from the original on 4 January 2007. Retrieved 3 January 2007.

39. Schombert, James (18 April 2006). "Elementary Particles". University of Oregon. Retrieved 3 January 2007.
40. Jevremovic, Tatjana (2005). Nuclear Principles in Engineering. Springer. p. 63. ISBN 0-387-23284-2. OCLC 228384008.
41. Pfeffer, Jeremy I.; Nir, Shlomo (2000). Modern Physics: An Introductory Text. Imperial College Press. pp. 330–336. ISBN 1-86094-250-4. OCLC 45900880.
42. Wenner, Jennifer M. (10 October 2007). "How Does Radioactive Decay Work?". Carleton College. Retrieved 9 January 2008.
43. Raymond, David (7 April 2006). "Nuclear Binding Energies". New Mexico Tech. Retrieved 3 January 2007.[dead link]
44. Mihos, Chris (23 July 2002). "Overcoming the Coulomb Barrier". Case Western Reserve University. Retrieved 13 February 2008.
45. Staff (30 March 2007). "ABC's of Nuclear Science". Lawrence Berkeley National Laboratory. Archived from the original on 5 December 2006. Retrieved 3 January 2007.
46. Makhijani, Arjun; Saleska, Scott (2 March 2001). "Basics of Nuclear Physics and Fission". Institute for Energy and Environmental Research. Archived from the original on 16 January 2007. Retrieved 3 January 2007.
47. Shultis, J. Kenneth; Faw, Richard E. (2002). Fundamentals of Nuclear Science and Engineering. CRC Press. pp. 10–17. ISBN 0-8247-0834-2. OCLC 123346507.
48. Fewell, M. P. (1995). "The atomic nuclide with the highest mean binding energy". American Journal of Physics. 63 (7): 653–658. Bibcode:1995AmJPh..63..653F. doi:10.1119/1.17828.
49. Mulliken, Robert S. (1967). "Spectroscopy, Molecular Orbitals, and Chemical Bonding". Science. 157 (3784): 13–24. Bibcode:1967Sci...157...13M. doi:10.1126/science.157.3784.13. PMID 5338306.
50. Brucat, Philip J. (2008). "The Quantum Atom". University of Florida. Archived from the original on 7 December 2006. Retrieved 4 January 2007.
51. Manthey, David (2001). "Atomic Orbitals". Orbital Central. Archived from the original on 10 January 2008. Retrieved 21 January 2008.
52. Herter, Terry (2006). "Lecture 8: The Hydrogen Atom". Cornell University. Archived from the original on 22 February 2012. Retrieved 14 February 2008.
53. Bell, R. E.; Elliott, L. G. (1950). "Gamma-Rays from the Reaction H1(n,γ)D2 and the Binding Energy of the Deuteron". Physical Review. 79 (2): 282–285. Bibcode:1950PhRv...79..282B. doi:10.1103/PhysRev.79.282.
54. Smirnov, Boris M. (2003). Physics of Atoms and Ions. Springer. pp. 249–272. ISBN 0-387-95550-X.
55. Matis, Howard S. (9 August 2000). "The Isotopes of Hydrogen". Guide to the Nuclear Wall Chart. Lawrence Berkeley National Lab. Archived from the original on 18 December 2007. Retrieved 21 December 2007.
56. Weiss, Rick (17 October 2006). "Scientists Announce Creation of Atomic

Element, the Heaviest Yet". Washington Post. Retrieved 21 December 2007.

57. Sills, Alan D. (2003). Earth Science the Easy Way. Barron's Educational Series. pp. 131–134. ISBN 0-7641-2146-4. OCLC 51543743.
58. Dumé, Belle (23 April 2003). "Bismuth breaks half-life record for alpha decay". Physics World. Archived from the original on 14 December 2007. Retrieved 21 December 2007.
59. Lindsay, Don (30 July 2000). "Radioactives Missing From The Earth". Don Lindsay Archive. Archived from the original on 28 April 2007. Retrieved 23 May 2007.
60. Tuli, Jagdish K. (April 2005). "Nuclear Wallet Cards". National Nuclear Data Center, Brookhaven National Laboratory. Retrieved 16 April 2011.
61. CRC Handbook (2002).
62. Mills, Ian; Cvitaš, Tomislav; Homann, Klaus; Kallay, Nikola; Kuchitsu, Kozo (1993). Quantities, Units and Symbols in Physical Chemistry (PDF) (2nd ed.). Oxford: International Union of Pure and Applied Chemistry, Commission on Physiochemical Symbols Terminology and Units, Blackwell Scientific Publications. p. 70. ISBN 0-632-03583-8. OCLC 27011505.
63. Chieh, Chung (22 January 2001). "Nuclide Stability". University of Waterloo. Retrieved 4 January 2007.
64. "Atomic Weights and Isotopic Compositions for All Elements". National Institute of Standards and Technology. Archived from the original on 31 December 2006. Retrieved 4 January 2007.
65. Audi, G.; Wapstra, A.H.; Thibault, C. (2003). "The Ame2003 atomic mass evaluation (II)" (PDF). Nuclear Physics A. 729 (1): 337–676. Bibcode:2003NuPhA.729..337A. doi:10.1016/j.nuclphysa.2003.11.003.
66. Shannon, R. D. (1976). "Revised effective ionic radii and systematic studies of interatomic distances in halides and chalcogenides". Acta Crystallographica A. 32 (5): 751–767. Bibcode:1976AcCrA..32..751S. doi:10.1107/S0567739476001551.
67. Dong, Judy (1998). "Diameter of an Atom". The Physics Factbook. Archived from the original on 4 November 2007. Retrieved 19 November 2007.
68. Zumdahl, Steven S. (2002). Introductory Chemistry: A Foundation (5th ed.). Houghton Mifflin. ISBN 0-618-34342-3. OCLC 173081482. Archived from the original on 4 March 2008. Retrieved 5 February 2008.
69. Birkholz, M.; Rudert, R. (2008). "Interatomic distances in pyrite-structure disulfides – a case for ellipsoidal modeling of sulfur ions]". phys. stat. sol. b. 245: 1858–1864. Bibcode:2008PSSBR.245.1858B. doi:10.1002/pssb.200879532.
70. Birkholz, M. (2014). "Modeling the Shape of Ions in Pyrite-Type Crystals". Crystals. 4: 390–403. doi:10.3390/cryst4030390Freely accessible.
71. Staff (2007). "Small Miracles: Harnessing nanotechnology". Oregon State University. Retrieved 7 January 2007.—describes the width of a human hair as 105 nm and 10 carbon atoms as spanning 1 nm.
72. Padilla, Michael J.; Miaoulis, Ioannis; Cyr, Martha (2002). Prentice Hall Science

Explorer: Chemical Building Blocks. Upper Saddle River, New Jersey USA: Prentice-Hall, Inc. p. 32. ISBN 0-13-054091-9. OCLC 47925884. There are 2,000,000,000,000,000,000,000 (that's 2 sextillion) atoms of oxygen in one drop of water—and twice as many atoms of hydrogen.

73. Feynman, Richard (1995). Six Easy Pieces. The Penguin Group. p. 5. ISBN 978-0-14-027666-4. OCLC 40499574.
74. "Radioactivity". Splung.com. Archived from the original on 4 December 2007. Retrieved 19 December 2007.
75. L'Annunziata, Michael F. (2003). Handbook of Radioactivity Analysis. Academic Press. pp. 3–56. ISBN 0-12-436603-1. OCLC 16212955.
76. Firestone, Richard B. (22 May 2000). "Radioactive Decay Modes". Berkeley Laboratory. Retrieved 7 January 2007.
77. Hornak, J. P. (2006). "Chapter 3: Spin Physics". The Basics of NMR. Rochester Institute of Technology. Archived from the original on 3 February 2007. Retrieved 7 January 2007.
78. Schroeder, Paul A. (25 February 2000). "Magnetic Properties". University of Georgia. Archived from the original on 29 April 2007. Retrieved 7 January 2007.
79. Goebel, Greg (1 September 2007). "[4.3] Magnetic Properties of the Atom". Elementary Quantum Physics. In The Public Domain website. Retrieved 7 January 2007.
80. Yarris, Lynn (Spring 1997). "Talking Pictures". Berkeley Lab Research Review. Archived from the original on 13 January 2008. Retrieved 9 January 2008.
81. Liang, Z.-P.; Haacke, E. M. (1999). Webster, J. G., ed. Encyclopedia of Electrical and Electronics Engineering: Magnetic Resonance Imaging. vol. 2. John Wiley & Sons. pp. 412–426. ISBN 0-471-13946-7.
82. Zeghbroeck, Bart J. Van (1998). "Energy levels". Shippensburg University. Retrieved 23 December 2007.[dead link]
83. Fowles, Grant R. (1989). Introduction to Modern Optics. Courier Dover Publications. pp. 227–233. ISBN 0-486-65957-7. OCLC 18834711.
84. Martin, W. C.; Wiese, W. L. (May 2007). "Atomic Spectroscopy: A Compendium of Basic Ideas, Notation, Data, and Formulas". National Institute of Standards and Technology. Archived from the original on 8 February 2007. Retrieved 8 January 2007.
85. "Atomic Emission Spectra — Origin of Spectral Lines". Avogadro Web Site. Retrieved 10 August 2006.
86. Fitzpatrick, Richard (16 February 2007). "Fine structure". University of Texas at Austin. Retrieved 14 February 2008.
87. Weiss, Michael (2001). "The Zeeman Effect". University of California-Riverside. Archived from the original on 2 February 2008. Retrieved 6 February 2008.
88. Beyer, H. F.; Shevelko, V. P. (2003). Introduction to the Physics of Highly Charged Ions. CRC Press. pp. 232–236. ISBN 0-7503-0481-2. OCLC 47150433.

89. Watkins, Thayer. "Coherence in Stimulated Emission". San José State University. Archived from the original on 12 January 2008. Retrieved 23 December 2007.
90. oxford dictionary – valency
91. Reusch, William (16 July 2007). "Virtual Textbook of Organic Chemistry". Michigan State University. Archived from the original on 29 October 2007. Retrieved 11 January 2008.
92. "Covalent bonding – Single bonds". chemguide. 2000.
93. Husted, Robert; et al. (11 December 2003). "Periodic Table of the Elements". Los Alamos National Laboratory. Archived from the original on 10 January 2008. Retrieved 11 January 2008.
94. Baum, Rudy (2003). "It's Elemental: The Periodic Table". Chemical & Engineering News. Retrieved 11 January 2008.
95. Goodstein, David L. (2002). States of Matter. Courier Dover Publications. pp. 436–438. ISBN 0-13-843557-X.
96. Brazhkin, Vadim V. (2006). "Metastable phases, phase transformations, and phase diagrams in physics and chemistry". Physics-Uspekhi. 49 (7): 719–24. Bibcode:2006PhyU...49..719B. doi:10.1070/PU2006v049n07ABEH006013.
97. Myers, Richard (2003). The Basics of Chemistry. Greenwood Press. p. 85. ISBN 0-313-31664-3. OCLC 50164580.
98. Staff (9 October 2001). "Bose-Einstein Condensate: A New Form of Matter". National Institute of Standards and Technology. Archived from the original on 3 January 2008. Retrieved 16 January 2008.
99. Colton, Imogen; Fyffe, Jeanette (3 February 1999). "Super Atoms from Bose-Einstein Condensation". The University of Melbourne. Archived from the original on 29 August 2007. Retrieved 6 February 2008.
100. Jacox, Marilyn; Gadzuk, J. William (November 1997). "Scanning Tunneling Microscope". National Institute of Standards and Technology. Archived from the original on 7 January 2008. Retrieved 11 January 2008.
101. "The Nobel Prize in Physics 1986". The Nobel Foundation. Retrieved 11 January 2008.—in particular, see the Nobel lecture by G. Binnig and H. Rohrer.
102. Jakubowski, N.; Moens, Luc; Vanhaecke, Frank (1998). "Sector field mass spectrometers in ICP-MS". Spectrochimica Acta Part B: Atomic Spectroscopy. 53 (13): 1739–63. Bibcode:1998AcSpe..53.1739J. doi:10.1016/S0584-8547(98)00222-5.
103. Müller, Erwin W.; Panitz, John A.; McLane, S. Brooks (1968). "The Atom-Probe Field Ion Microscope". Review of Scientific Instruments. 39 (1): 83–86. Bibcode:1968RScI...39...83M. doi:10.1063/1.1683116.
104. Lochner, Jim; Gibb, Meredith; Newman, Phil (30 April 2007). "What Do Spectra Tell Us?". NASA/Goddard Space Flight Center. Archived from the original on 16 January 2008. Retrieved 3 January 2008.
105. Winter, Mark (2007). "Helium". WebElements. Archived from the original on 30 December 2007. Retrieved 3 January 2008.

106. Hinshaw, Gary (10 February 2006). "What is the Universe Made Of?". NASA/WMAP. Archived from the original on 31 December 2007. Retrieved 7 January 2008.

107. Choppin, Gregory R.; Liljenzin, Jan-Olov; Rydberg, Jan (2001). Radiochemistry and Nuclear Chemistry. Elsevier. p. 441. ISBN 0-7506-7463-6. OCLC 162592180.

108. Davidsen, Arthur F. (1993). "Far-Ultraviolet Astronomy on the Astro-1 Space Shuttle Mission". Science. 259 (5093): 327–34. Bibcode:1993Sci...259..327D. doi:10.1126/science.259.5093.327. PMID 17832344.

109. Lequeux, James (2005). The Interstellar Medium. Springer. p. 4. ISBN 3-540-21326-0. OCLC 133157789.

110. Smith, Nigel (6 January 2000). "The search for dark matter". Physics World. Archived from the original on 16 February 2008. Retrieved 14 February 2008.

111. Croswell, Ken (1991). "Boron, bumps and the Big Bang: Was matter spread evenly when the Universe began? Perhaps not; the clues lie in the creation of the lighter elements such as boron and beryllium". New Scientist (1794): 42. Archived from the original on 7 February 2008. Retrieved 14 January 2008.

112. Copi, Craig J.; Schramm, DN; Turner, MS (1995). "Big-Bang Nucleosynthesis and the Baryon Density of the Universe". Science. 267 (5195): 192–99. arXiv:astro-ph/9407006Freely accessible. Bibcode:1995Sci...267..192C. doi:10.1126/science.7809624. PMID 7809624.

113. Hinshaw, Gary (15 December 2005). "Tests of the Big Bang: The Light Elements". NASA/WMAP. Archived from the original on 17 January 2008. Retrieved 13 January 2008.

114. Abbott, Brian (30 May 2007). "Microwave (WMAP) All-Sky Survey". Hayden Planetarium. Archived from the original on 13 February 2013. Retrieved 13 January 2008.

115. Hoyle, F. (1946). "The synthesis of the elements from hydrogen". Monthly Notices of the Royal Astronomical Society. 106: 343–83. Bibcode:1946MNRAS.106..343H. doi:10.1093/mnras/106.5.343.

116. Knauth, D. C.; Knauth, D. C.; Lambert, David L.; Crane, P. (2000). "Newly synthesized lithium in the interstellar medium". Nature. 405 (6787): 656–58. doi:10.1038/35015028. PMID 10864316.

117. Mashnik, Stepan G. (2000). "On Solar System and Cosmic Rays Nucleosynthesis and Spallation Processes". arXiv:astro-ph/0008382Freely accessible [astro-ph].

118. Kansas Geological Survey (4 May 2005). "Age of the Earth". University of Kansas. Retrieved 14 January 2008.

119. Manuel 2001, pp. 407–430, 511–519.

120. Dalrymple, G. Brent (2001). "The age of the Earth in the twentieth century: a problem (mostly) solved". Geological Society, London, Special Publications. 190 (1): 205–21. Bibcode:2001GSLSP.190..205D. doi:10.1144/GSL.SP.2001.190.01.14. Retrieved 14 January 2008.

121. Anderson, Don L.; Foulger, G. R.; Meibom, Anders (2 September 2006).

"Helium: Fundamental models". MantlePlumes.org. Archived from the original on 8 February 2007. Retrieved 14 January 2007.

122. Pennicott, Katie (10 May 2001). "Carbon clock could show the wrong time". PhysicsWeb. Archived from the original on 15 December 2007. Retrieved 14 January 2008.
123. Yarris, Lynn (27 July 2001). "New Superheavy Elements 118 and 116 Discovered at Berkeley Lab". Berkeley Lab. Archived from the original on 9 January 2008. Retrieved 14 January 2008.
124. Diamond, H; et al. (1960). "Heavy Isotope Abundances in Mike Thermonuclear Device". Physical Review. 119 (6): 2000–04. Bibcode:1960PhRv..119.2000D. doi:10.1103/PhysRev.119.2000.
125. Poston Sr.; John W. (23 March 1998). "Do transuranic elements such as plutonium ever occur naturally?". Scientific American.
126. Keller, C. (1973). "Natural occurrence of lanthanides, actinides, and superheavy elements". Chemiker Zeitung. 97 (10): 522–30. OSTI 4353086.
127. Zaider, Marco; Rossi, Harald H. (2001). Radiation Science for Physicians and Public Health Workers. Springer. p. 17. ISBN 0-306-46403-9. OCLC 44110319.
128. Manuel 2001, pp. 407–430,511–519.
129. "Oklo Fossil Reactors". Curtin University of Technology. Archived from the original on 18 December 2007. Retrieved 15 January 2008.
130. Weisenberger, Drew. "How many atoms are there in the world?". Jefferson Lab. Retrieved 16 January 2008.
131. Pidwirny, Michael. "Fundamentals of Physical Geography". University of British Columbia Okanagan. Archived from the original on 21 January 2008. Retrieved 16 January 2008.
132. Anderson, Don L. (2002). "The inner inner core of Earth". Proceedings of the National Academy of Sciences. 99 (22): 13966–68. Bibcode:2002PNAS...9913966A. doi:10.1073/pnas.232565899. PMC 137819Freely accessible. PMID 12391308.
133. Pauling, Linus (1960). The Nature of the Chemical Bond. Cornell University Press. pp. 5–10. ISBN 0-8014-0333-2. OCLC 17518275.
134. Anonymous (2 October 2001). "Second postcard from the island of stability". CERN Courier. Archived from the original on 3 February 2008. Retrieved 14 January 2008.
135. Jacoby, Mitch (2006). "As-yet-unsynthesized superheavy atom should form a stable diatomic molecule with fluorine". Chemical & Engineering News. 84 (10): 19. doi:10.1021/cen-v084n010.p019a.
136. Koppes, Steve (1 March 1999). "Fermilab Physicists Find New Matter-Antimatter Asymmetry". University of Chicago. Retrieved 14 January 2008.
137. Cromie, William J. (16 August 2001). "A lifetime of trillionths of a second: Scientists explore antimatter". Harvard University Gazette. Retrieved 14 January 2008.

138. Hijmans, Tom W. (2002). "Particle physics: Cold antihydrogen". Nature. 419 (6906): 439–40. Bibcode:2002Natur.419..439H. doi:10.1038/419439aFreely accessible. PMID 12368837.

139. Staff (30 October 2002). "Researchers ‹look inside' antimatter". BBC News. Retrieved 14 January 2008.

140. Barrett, Roger (1990). "The Strange World of the Exotic Atom". New Scientist (1728): 77–115. Archived from the original on 21 December 2007. Retrieved 4 January 2008.

141. Indelicato, Paul (2004). "Exotic Atoms". Physica Scripta. T112 (1): 20–26. arXiv:physics/0409058Freely accessible. Bibcode:2004PhST..112...20I. doi:10.1238/Physica.Topical.112a00020.

142. Ripin, Barrett H. (July 1998). "Recent Experiments on Exotic Atoms". American Physical Society. Retrieved 15 February 2008.

CHAPTER 3

Fermions

Chapter Contents

Fermions are particles which have half-integer spin and therefore are constrained by the Pauli exclusion principle. Particles with integer spin are called bosons. Fermions include electrons, protons, neutrons. The wavefunction which describes a collection of fermions must be antisymmetric with respect to the exchange of identical particles, while the wavefunction for a collection of bosons is symmetric.

INTRODUCTION

In particle physics, a fermion (a name coined by Paul Dirac from the surname of Enrico Fermi) is any particle characterized by Fermi–Dirac statistics. These particles obey the Pauli exclusion principle. Fermions include all quarks and leptons, as well as any composite particle made of an odd number of these, such as all baryons and many atoms and nuclei. Fermions differ from bosons, which obey Bose–Einstein statistics.

A fermion can be an elementary particle, such as the electron, or it can be a composite particle, such as the proton. According to the spin-statistics theorem in any reasonable relativistic quantum field theory, particles with integer spin are bosons, while particles with half-integer spin are fermions. Besides this spin characteristic, fermions have another specific property: they possess conserved baryon or lepton quantum numbers. Therefore, what is usually referred to as the spin statistics relation is in fact a spin statistics-quantum number relation.

Enrico Fermi was an Italian physicist, who created the world's first nuclear reactor, the Chicago Pile-1. He has been called the "architect of the nuclear age"

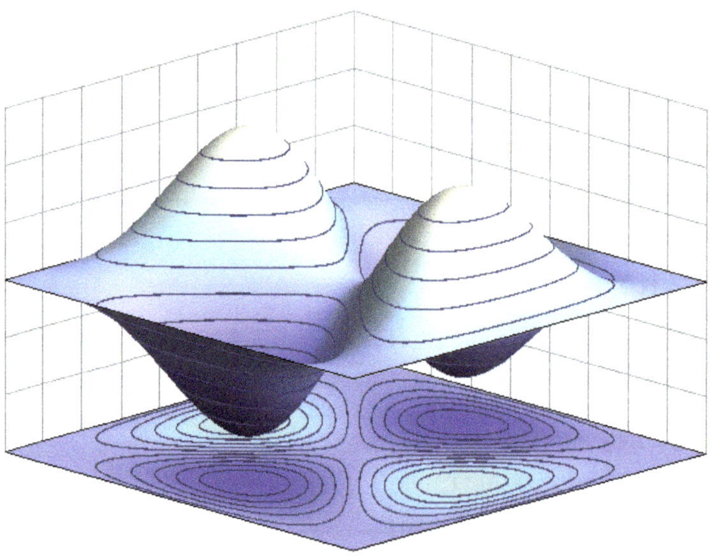

Antisymmetric wavefunction for a (fermionic) 2-particle state in an infinite square well potential.

As a consequence of the Pauli exclusion principle, only one fermion can occupy a particular quantum state at any given time. If multiple fermions have the same spatial probability distribution, then at least one property of each fermion, such as its spin, must be different. Fermions are usually associated with matter, whereas bosons are generally force carrier particles, although in the current state of particle physics the distinction between the two concepts is unclear. Weakly interacting fermions can also display bosonic behavior under extreme conditions. At low temperature fermions show superfluidity for uncharged particles and superconductivity for charged particles.

Composite fermions, such as protons and neutrons, are the key building blocks of everyday matter.

ELEMENTARY FERMIONS

The Standard Model recognizes two types of elementary fermions: quarks and leptons. In all, the model distinguishes 24 different fermions. There are six quarks (up, down, strange, charm, bottom and top quarks), and six leptons (electron, electron neutrino, muon, muon neutrino, tau particle and tau neutrino), along with the corresponding antiparticle of each of these.

Mathematically, fermions come in three types - Weyl fermions (massless), Dirac fermions (massive), and Majorana fermions (each its own antiparticle). Most Standard Model fermions are believed to be Dirac fermions, although it is unknown at this time whether the neutrinos are Dirac or Majorana fermions. Dirac fermions can be treated as a combination of two Weyl fermions. In July 2015, Weyl fermions have been experimentally realized in Weyl semimetals.

COMPOSITE FERMIONS

Composite particles (such as hadrons, nuclei, and atoms) can be bosons or fermions depending on their constituents. More precisely, because of the relation between spin and statistics, a particle containing an odd number of fermions is itself a fermion. It will have half-integer spin.

Examples include the following:

- A baryon, such as the proton or neutron, contains three fermionic quarks and thus it is a fermion.
- The nucleus of a carbon-13 atom contains six protons and seven neutrons and is therefore a fermion.
- The atom helium-3 (^3He) is made of two protons, one neutron, and two electrons, and therefore it is a fermion.

The number of bosons within a composite particle made up of simple particles bound with a potential has no effect on whether it is a boson or a fermion.

Fermionic or bosonic behavior of a composite particle (or system) is only seen at large (compared to size of the system) distances. At proximity, where spatial structure begins to be important, a composite particle (or system) behaves according to its constituent makeup.

Fermions can exhibit bosonic behavior when they become loosely bound in pairs. This is the origin of superconductivity and the superfluidity of helium-3: in superconducting materials, electrons interact through the exchange of phonons, forming Cooper pairs, while in helium-3, Cooper pairs are formed via spin fluctuations.

The quasiparticles of the fractional quantum Hall effect are also known as composite fermions, which are electrons with an even number of quantized vortices attached to them.

Skyrmions

In a quantum field theory, there can be field configurations of bosons which are topologically twisted. These are coherent states (or solitons) which behave like a particle, and they can be fermionic even if all the constituent particles are bosons. This was discovered by Tony Skyrme in the early 1960s, so fermions made of bosons are named skyrmions after him.

Skyrme's original example involved fields which take values on a three-dimensional sphere, the original nonlinear sigma model which describes the large distance behavior of pions. In Skyrme's model, reproduced in the large N or string approximation to quantum chromodynamics (QCD), the proton and neutron are fermionic topological solitons of the pion field.

Whereas Skyrme's example involved pion physics, there is a much more familiar example in quantum electrodynamics with a magnetic monopole. A bosonic monopole with the smallest possible magnetic charge and a bosonic version of the electron will form a fermionic dyon.

The analogy between the Skyrme field and the Higgs field of the electroweak sector has been used to postulate that all fermions are skyrmions. This could explain why all known fermions have baryon or lepton quantum numbers and provide a physical mechanism for the Pauli exclusion principle.

QUARK

A quark is an elementary particle and a fundamental constituent of matter. Quarks combine to form composite particles called hadrons, the most stable of which are protons and neutrons, the components of atomic nuclei. Due to a phenomenon known as color confinement, quarks are never directly observed or found in isolation; they can be found only within hadrons, such as baryons (of which protons and neutrons are examples) and mesons. For this reason, much of what is known about quarks has been drawn from observations of the hadrons themselves.

Quarks have various intrinsic properties, including electric charge, mass, color charge, and spin. Quarks are the only elementary particles in the Standard Model of particle physics to experience all four fundamental interactions, also known as fundamental forces (electromagnetism, gravitation, strong interaction, and weak interaction), as well as the only known particles whose electric charges are not integer multiples of the elementary charge.

There are six types of quarks, known as flavors: up, down, strange, charm, top, and bottom. Up and down quarks have the lowest masses of all quarks. The heavier quarks rapidly change into up and down quarks through a process of particle decay: the transformation from a higher mass state to a lower mass state. Because of this, up and down quarks are generally stable and the most common in the universe, whereas strange, charm, bottom, and top quarks can only be produced in high energy collisions (such as those involving cosmic rays and in particle accelerators). For every quark flavor there is a corresponding type of antiparticle, known as an antiquark, that differs from the quark only in that some of its properties have equal magnitude but opposite sign.

The quark model was independently proposed by physicists Murray Gell-Mann and George Zweig in 1964. Quarks were introduced as parts of an ordering scheme for hadrons, and there was little evidence for their physical existence until deep inelastic scattering experiments at the Stanford Linear Accelerator Center in 1968. Accelerator experiments have provided evidence for all six flavors. The top quark was the last to be discovered at Fermilab in 1995.

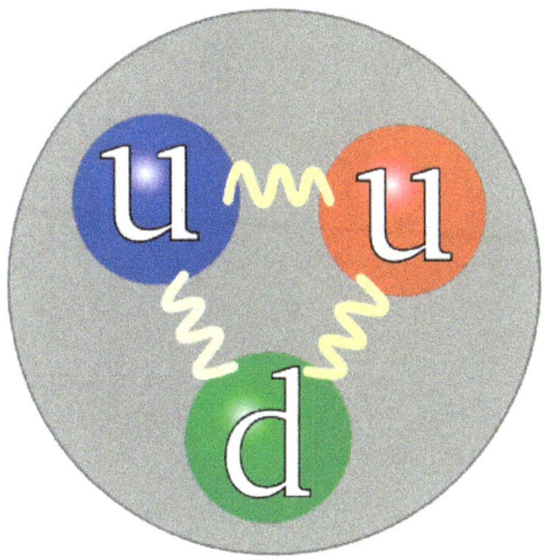

A proton is composed of two up quarks, one down quark, and the gluons that mediate the forces "binding" them together. The color assignment of individual quarks is arbitrary, but all three colors must be present.

Classification

The Standard Model is the theoretical framework describing all the currently known elementary particles. This model contains six flavors of quarks (q), named up (u), down (d), strange (s), charm (c), bottom (b), and top (t). Antiparticles of quarks are called antiquarks, and are denoted by a bar over the symbol for the corresponding quark, such as u for an up antiquark. As with antimatter in general, antiquarks have the same mass, mean lifetime, and spin as their respective quarks, but the electric charge and other charges have the opposite sign.

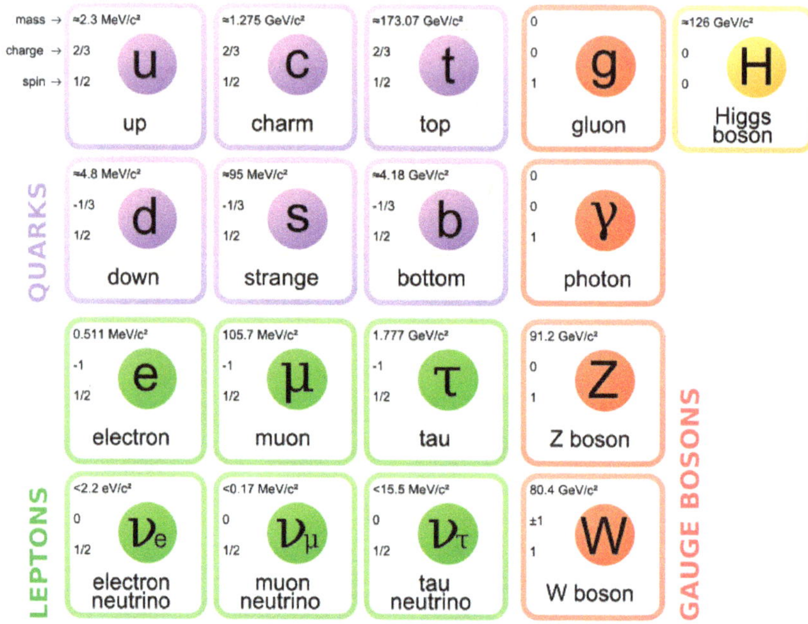

Six of the particles in the Standard Model are quarks (shown in purple). Each of the first three columns forms a generation of matter.

Quarks are spin-$\frac{1}{2}$ particles, implying that they are fermions according to the spin-statistics theorem. They are subject to the Pauli exclusion principle, which states that no two identical fermions can simultaneously occupy the same quantum state. This is in contrast to bosons (particles with integer spin), any number of which can be in the same state.Unlike leptons, quarks possess color charge, which causes them to engage in the strong interaction. The resulting attraction between different quarks causes the formation of composite particles known as hadrons.

The quarks that determine the quantum numbers of hadrons are called valence quarks; apart from these, any hadron may contain an indefinite number of virtual (or sea) quarks, antiquarks, and gluons, which do not influence its quantum numbers. There are two families of hadrons: baryons, with three valence quarks, and mesons, with a valence quark and an antiquark. The most common baryons are the proton and the neutron, the building blocks of the atomic nucleus. A great number of hadrons are known, most of them differentiated by their quark content and the properties these constituent quarks confer. The existence of "exotic" hadrons with more valence quarks, such as tetraquarks (qqqq) and pentaquarks (qqqqq), has been conjectured but not proven. However, on 13 July 2015, the LHCb collaboration at CERN reported results consistent with pentaquark states.

Elementary fermions are grouped into three generations, each comprising two leptons and two quarks. The first generation includes up and down quarks, the second strange and charm quarks, and the third bottom and top quarks. All searches for a fourth generation of quarks and other elementary fermions have failed, and there is strong indirect evidence that no more than three generations exist. Particles in higher generations generally have greater mass and less stability, causing them to decay into lower-generation particles by means of weak interactions. Only first-generation (up and down) quarks occur commonly in nature. Heavier quarks can only be created in high-energy collisions (such as in those involving cosmic rays), and decay quickly; however, they are thought to have been present during the first fractions of a second after the Big Bang, when the universe was in an extremely hot and dense phase (the quark epoch). Studies of heavier quarks are conducted in artificially created conditions, such as in particle accelerators.

Having electric charge, mass, color charge, and flavor, quarks are the only known elementary particles that engage in all four fundamental interactions of contemporary physics: electromagnetism, gravitation, strong interaction, and weak interaction. Gravitation is too weak to be relevant to individual particle interactions except at extremes of energy (Planck energy) and distance scales (Planck distance). However, since no successful quantum theory of gravity exists, gravitation is not described by the Standard Model.

History

The quark model was independently proposed by physicists Murray Gell-Mann (pictured) and George Zweig in 1964. The proposal came shortly after Gell-Mann's 1961 formulation of a particle classification system known as the Eightfold Way—or, in more technical terms, SU(3) flavor symmetry, streamlining its structure. Physicist Yuval Ne'eman had independently developed a scheme similar to the Eightfold Way in the same year. An early attempt at constituent organization was available in the Sakata model.

Murray Gell-Mann at TED in 2007. Gell-Mann and George Zweig proposed the quark model in 1964.

At the time of the quark theory's inception, the "particle zoo" included, amongst other particles, a multitude of hadrons. Gell-Mann and Zweig posited that they were not elementary particles, but were instead composed of combinations of quarks and antiquarks. Their model involved three flavors of quarks, up, down, and strange, to which they ascribed properties such as spin and electric charge. The initial reaction of the physics community to the proposal was mixed. There was particular contention about whether the quark was a physical entity or a mere abstraction used to explain concepts that were not fully understood at the time.

In less than a year, extensions to the Gell-Mann–Zweig model were proposed. Sheldon Lee Glashow and James Bjorken predicted the existence of a fourth flavor of quark, which they called charm. The addition was proposed because it allowed for a better description of the weak interaction (the mechanism that allows quarks to decay), equalized the number of known quarks with the number of known leptons, and implied a mass formula that correctly reproduced the masses of the known mesons.

In 1968, deep inelastic scattering experiments at the Stanford Linear Accelerator Center (SLAC) showed that the proton contained much smaller, point-like objects and was therefore not an elementary particle. Physicists were reluctant to firmly identify these objects with quarks at the time, instead calling them "partons"—a term coined by Richard Feynman. The objects that were observed at SLAC would later be identified as up and down quarks as the other flavors were discovered. Nevertheless, "parton" remains in use as a collective term for the constituents of hadrons (quarks, antiquarks, and gluons).

The strange quark's existence was indirectly validated by SLAC's scattering experiments: not only was it a necessary component of Gell-Mann and Zweig's three-quark model, but it provided an explanation for the kaon (K) and pion (π) hadrons discovered in cosmic rays in 1947.

In a 1970 paper, Glashow, John Iliopoulos and Luciano Maiani presented the so-called GIM mechanism to explain the experimental non-observation of flavor-changing neutral currents. This theoretical model required the existence of the as-yet undiscovered charm quark. The number of supposed quark flavors grew to the current six in 1973, when Makoto Kobayashi and Toshihide Maskawa noted that the experimental observation of CP violation could be explained if there were another pair of quarks.

Charm quarks were produced almost simultaneously by two teams in November 1974—one at SLAC under Burton Richter, and one at Brookhaven National Laboratory under Samuel Ting. The charm quarks were observed bound with charm antiquarks in mesons. The two parties had assigned the discovered meson two different symbols, J and ψ; thus, it became formally known as the J/ψ meson. The discovery finally convinced the physics community of the quark model›s validity.

In the following years a number of suggestions appeared for extending the quark model to six quarks. Of these, the 1975 paper by Haim Harari was the first to coin the terms top and bottom for the additional quarks.

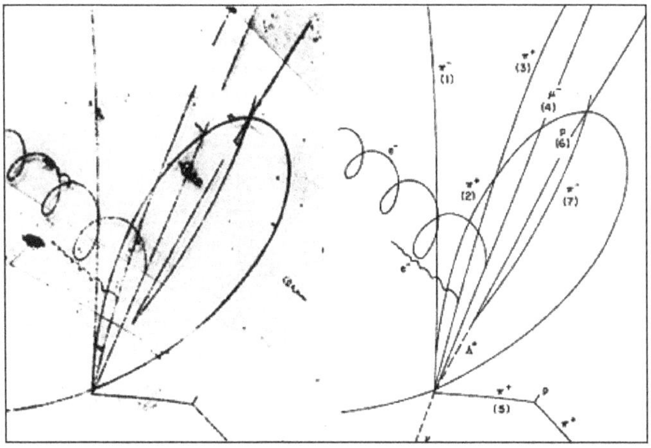

Photograph of the event that led to the discovery of the Σ++c baryon, at the Brookhaven National Laboratory in 1974

In 1977, the bottom quark was observed by a team at Fermilab led by Leon Lederman. This was a strong indicator of the top quark's existence: without the top quark, the bottom quark would have been without a partner. However, it was not until 1995 that the top quark was finally observed, also by the CDF and DØ teams at Fermilab. It had a mass much larger than had been previously expected, almost as large as that of a gold atom.

Etymology

For some time, Gell-Mann was undecided on an actual spelling for the term he intended to coin, until he found the word quark in James Joyce's book Finnegans Wake:

--Three quarks for Muster Mark!Sure he hasn't got much of a barkAnd sure any he has it's all beside the mark.

—James Joyce, Finnegans Wake

Gell-Mann went into further detail regarding the name of the quark in his book The Quark and the Jaguar:

In 1963, when I assigned the name "quark" to the fundamental constituents of the nucleon, I had the sound first, without the spelling, which could have been "kwork". Then, in one of my occasional perusals of Finnegans Wake, by James Joyce, I came across the word "quark" in the phrase "Three quarks for Muster Mark". Since "quark" (meaning, for one thing, the cry of the gull) was clearly intended to rhyme with "Mark", as well as "bark" and other such words, I had to find an excuse to pronounce it as "kwork". But the book represents the dream of a publican named Humphrey Chimpden Earwicker. Words in the text are typically drawn from several sources at once, like the "portmanteau" words in Through the Looking-Glass. From time to time, phrases occur in the book that are partially determined by calls for drinks at the bar. I argued, therefore, that perhaps one of the multiple sources of the cry "Three quarks for Muster Mark" might be "Three quarts for Mister Mark", in which case the pronunciation "kwork" would not be totally unjustified. In any case, the number three fitted perfectly the way quarks occur in nature.

Zweig preferred the name ace for the particle he had theorized, but Gell-Mann's terminology came to prominence once the quark model had been commonly accepted.

The quark flavors were given their names for several reasons. The up and down quarks are named after the up and down components of isospin, which they carry. Strange quarks were given their name because they were discovered to be components of the strange particles discovered in cosmic rays years before the quark model was proposed; these particles were deemed "strange" because they had unusually long lifetimes. Glashow, who coproposed charm quark with Bjorken, is quoted as saying, "We called our construct the 'charmed quark', for we were fascinated and pleased by the symmetry it brought to the subnuclear world." The names "bottom" and "top", coined by Harari, were chosen because they are "logical partners for up and down quarks". In the past, bottom and top quarks were sometimes referred to as "beauty" and "truth" respectively, but these names have somewhat fallen out of use. While "truth" never did catch on, accelerator complexes devoted to massive production of bottom quarks are sometimes called "beauty factories".

Properties

Electric Charge

Quarks have fractional electric charge values – either $1/3$ or $2/3$ times the elementary charge (e), depending on flavor. Up, charm, and top quarks (collectively referred to as up-type quarks) have a charge of $+2/3$ e, while down, strange, and bottom quarks (down-type quarks) have $-1/3$ e. Antiquarks have the opposite charge to their corresponding quarks; up-type antiquarks have charges of $-2/3$ e and down-type antiquarks have charges of $+1/3$ e. Since the electric charge of a hadron is the sum of the charges of the constituent quarks, all hadrons have integer charges: the combination of three quarks (baryons), three antiquarks (antibaryons), or a quark and an antiquark (mesons) always results in integer charges. For example, the hadron constituents of atomic nuclei, neutrons and protons, have charges of 0 e and +1 e respectively; the neutron is composed of two down quarks and one up quark, and the proton of two up quarks and one down quark.

Spin

Spin is an intrinsic property of elementary particles, and its direction is an important degree of freedom. It is sometimes visualized as the rotation of an object around its own axis (hence the name "spin"), though this notion is somewhat misguided at subatomic scales because elementary particles are believed to be point-like.

Spin can be represented by a vector whose length is measured in units of the reduced Planck constant ħ (pronounced "h bar"). For quarks, a measurement of the spin vector component along any axis can only yield the values $+\hbar/2$ or $-\hbar/2$; for this reason quarks are classified as spin-$1/2$ particles. The component of spin along a given axis – by convention the z axis – is often denoted by an up arrow ↑ for the value $+1/2$ and down arrow ↓ for the value $-\frac{1}{2}$, placed after the symbol for flavor. For example, an up quark with a spin of $+\frac{1}{2}$ along the z axis is denoted by u↑.

Weak Interaction

A quark of one flavor can transform into a quark of another flavor only through the weak interaction, one of the four fundamental interactions in particle physics. By absorbing or emitting a W boson, any up-type quark (up, charm, and top quarks) can change into any down-type quark (down, strange, and bottom quarks) and vice versa. This flavor transformation mechanism causes the radioactive process of beta decay, in which a neutron (n) "splits" into a proton (p), an electron (e−) and an electron antineutrino (ve) (see picture).

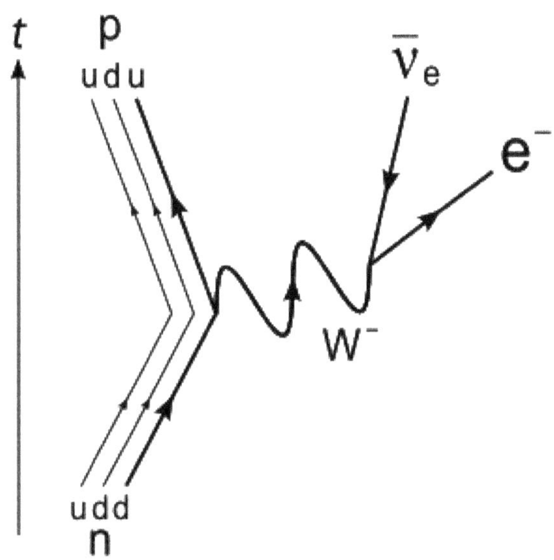

Feynman diagram of beta decay with time flowing upwards. The CKM matrix (discussed below) encodes the probability of this and other quark decays.

This occurs when one of the down quarks in the neutron (udd) decays into an up quark by emitting a virtual W− boson, transforming the neutron into a proton (uud). The W− boson then decays into an electron and an electron antineutrino.

$n \rightarrow p + e^- + \bar{\nu}_e$ (Bete decay, hadron notation)

$udd \rightarrow uud + e^- + \bar{\nu}_e$ (Bete decay, hadron notation)

Both beta decay and the inverse process of inverse beta decay are routinely used in medical applications such as positron emission tomography (PET) and in experiments involving neutrino detection.

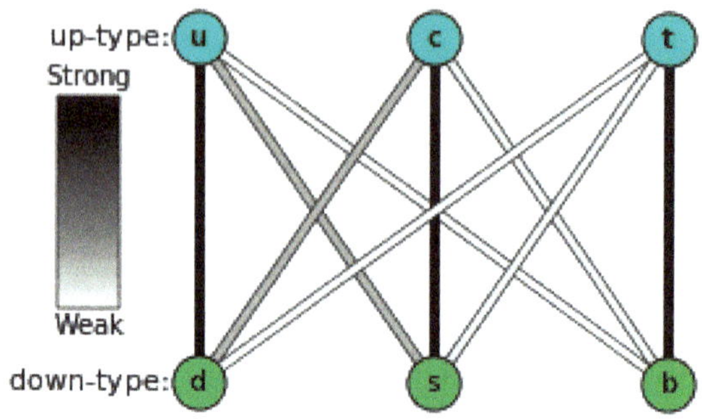

The strengths of the weak interactions between the six quarks. The "intensities" of the lines are determined by the elements of the CKM matrix.

While the process of flavor transformation is the same for all quarks, each quark has a preference to transform into the quark of its own generation. The relative tendencies of all flavor transformations are described by a mathematical table, called the Cabibbo–Kobayashi–Maskawa matrix (CKM matrix). Enforcing unitarity, the approximate magnitudes of the entries of the CKM matrix are:

$$\begin{bmatrix} |V_{\mathrm{ud}}| & |V_{\mathrm{us}}| & |V_{\mathrm{ub}}| \\ |V_{\mathrm{cd}}| & |V_{\mathrm{cs}}| & |V_{\mathrm{cb}}| \\ |V_{\mathrm{td}}| & |V_{\mathrm{ts}}| & |V_{\mathrm{tb}}| \end{bmatrix} \approx \begin{bmatrix} 0.974 & 0.225 & 0.003 \\ 0.225 & 0.973 & 0.041 \\ 0.009 & 0.040 & 0.999 \end{bmatrix},$$

where V_{ij} represents the tendency of a quark of flavor i to change into a quark of flavor j (or vice versa).

There exists an equivalent weak interaction matrix for leptons (right side of the W boson on the above beta decay diagram), called the Pontecorvo–Maki–Nakagawa–Sakata matrix (PMNS matrix). Together, the CKM and PMNS matrices describe all flavor transformations, but the links between the two are not yet clear.

Strong Interaction and Color Charge

According to quantum chromodynamics (QCD), quarks possess a property called color charge. There are three types of color charge, arbitrarily labeled blue, green, and red. Each of them is complemented by an anticolor – antiblue, antigreen, and antired. Every quark carries a color, while every antiquark carries an anticolor.

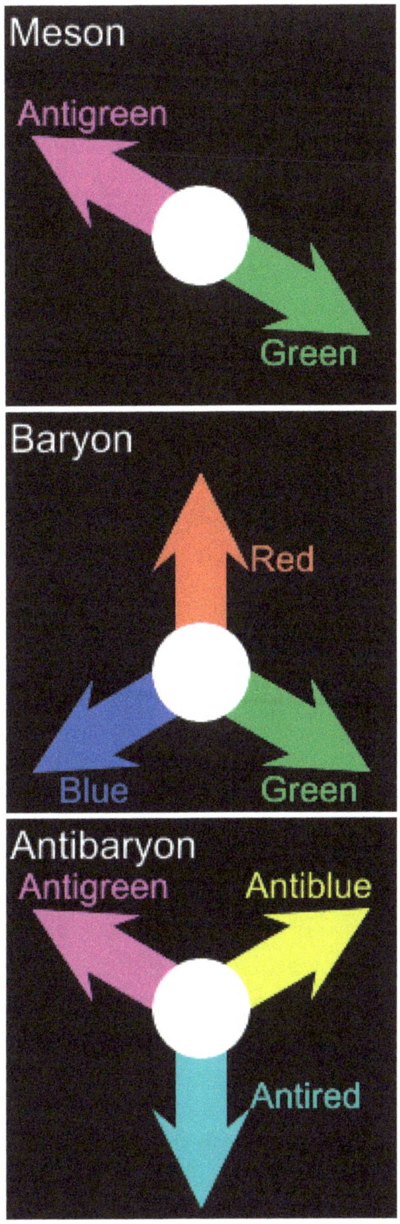

All types of hadrons have zero total color charge.

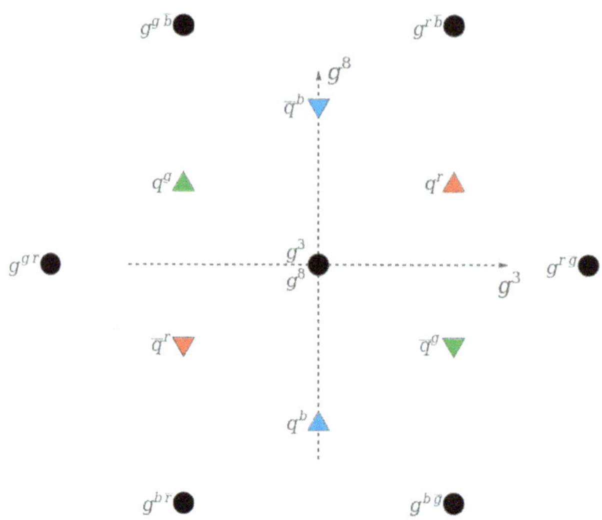

The pattern of strong charges for the three colors of quark, three antiquarks, and eight gluons (with two of zero charge overlapping).

The system of attraction and repulsion between quarks charged with different combinations of the three colors is called strong interaction, which is mediated by force carrying particles known as gluons; this is discussed at length below. The theory that describes strong interactions is called quantum chromodynamics (QCD). A quark, which will have a single color value, can form a bound system with an antiquark carrying the corresponding anticolor. The result of two attracting quarks will be color neutrality: a quark with color charge ξ plus an antiquark with color charge −ξ will result in a color charge of 0 (or "white" color) and the formation of a meson. This is analogous to the additive color model in basic optics. Similarly, the combination of three quarks, each with different color charges, or three antiquarks, each with anticolor charges, will result in the same "white" color charge and the formation of a baryon or antibaryon.

In modern particle physics, gauge symmetries – a kind of symmetry group – relate interactions between particles. Color SU(3) (commonly abbreviated to $SU(3)_c$) is the gauge symmetry that relates the color charge in quarks and is the defining symmetry for quantum chromodynamics. Just as the laws of physics are independent of which directions in space are designated x, y, and z, and remain unchanged if the coordinate axes are rotated to a new orientation, the physics of quantum chromodynamics is independent of which directions in three-dimensional color space are identified as blue, red, and green. $SU(3)_c$ color transformations correspond to "rotations" in color space (which, mathematically speaking, is a complex space). Every quark flavor f, each with subtypes f_B, f_G, $f_R$corresponding to the quark colors, forms a triplet: a three-component quantum field which transforms under the fundamental representation of $SU(3)_c$. The requirement that $SU(3)_c$ should be local – that is, that its transformations be allowed to vary with space and time – determines the properties of the strong interaction. In particular, it implies the existence of eight gluon types to act as its force carriers.

Mass

Two terms are used in referring to a quark's mass: current quark mass refers to the mass of a quark by itself, while constituent quark mass refers to the current quark mass plus the mass of the gluonparticle field surrounding the quark. These masses typically have very different values. Most of a hadron's mass comes from the gluons that bind the constituent quarks together, rather than from the quarks themselves. While gluons are inherently massless, they possess energy – more specifically, quantum chromodynamics binding energy (QCBE) – and it is this that contributes so greatly to the overall mass of the hadron. For example, a proton has a mass of approximately 938 MeV/c^2, of which the rest mass of its three valence quarks only contributes about 9 MeV/c^2; much of the remainder can be attributed to the field energy of the gluons.

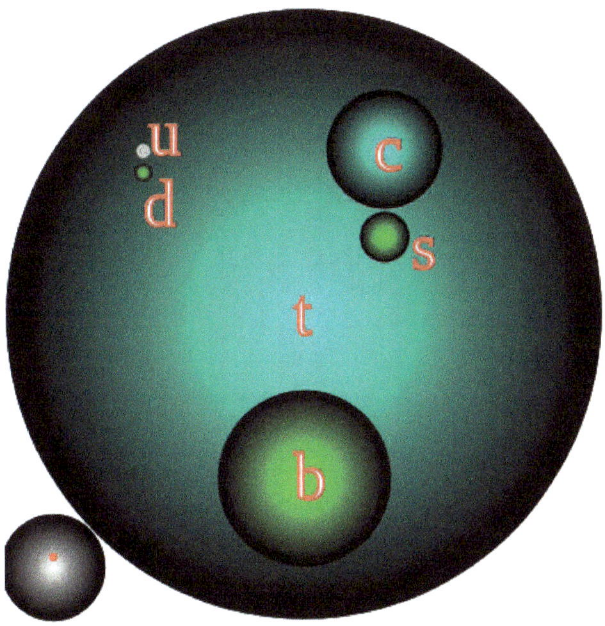

Current quark masses for all six flavors in comparison, as balls of proportional volumes. Proton and electron (red) are shown in bottom left corner for scale

The Standard Model posits that elementary particles derive their masses from the Higgs mechanism, which is associated to the Higgs boson. It is hoped that further research into the reasons for the top quark's large mass of ~173 GeV/c^2, almost the mass of a gold atom, might reveal more about the origin of the mass of quarks and other elementary particles.

Table of Properties

The following table summarizes the key properties of the six quarks. Flavor quantum numbers (isospin (I_3), charm (C), strangeness (S, not to be confused with spin), topness (T), and bottomness (B′)) are assigned to certain quark flavors, and denote qualities of quark-based systems and hadrons. The baryon number (B) is $+\frac{1}{3}$ for all quarks, as baryons are made of three quarks. For antiquarks, the electric charge (Q) and all flavor quantum numbers (B, I_3, C, S, T, and B′) are of opposite sign. Mass and total angular momentum (J; equal to spin for point particles) do not change sign for the antiquarks.

Quark flavor properties												
Name	**Symbol**	**Mass (MeV/c^2)***	***J***	***B***	***Q* (e)**	***I*₃**	***C***	***S***	***T***	***B′***	**Antiparticle**	**Antiparticle symbol**
First generation												
Up	u	2.3±0.7 ± 0.5	$\frac{1}{2}$	$+\frac{1}{3}$	$+\frac{2}{3}$	$+\frac{1}{2}$	0	0	0	0	Antiup	u
Down	d	4.8±0.5 ± 0.3	$\frac{1}{2}$	$+\frac{1}{3}$	$-\frac{1}{3}$	$-\frac{1}{2}$	0	0	0	0	Antidown	d
Second generation												
Charm	c	1275±25	$\frac{1}{2}$	$+\frac{1}{3}$	$+\frac{2}{3}$	0	+1	0	0	0	Anticharm	c
Strange	s	95±5	$\frac{1}{2}$	$+\frac{1}{3}$	$-\frac{1}{3}$	0	0	−1	0	0	Antistrange	s
Third generation												
Top	t	173210±510 ± 710	$\frac{1}{2}$	$+\frac{1}{3}$	$+\frac{2}{3}$	0	0	0	+1	0	Antitop	t
Bottom	b	4180±30	$\frac{1}{2}$	$+\frac{1}{3}$	$-\frac{1}{3}$	0	0	0	0	−1	Antibottom	b

J = total angular momentum, B = baryon number, Q = electric charge, I_3 = isospin, C = charm, S = strangeness, T = topness, B′ = bottomness.* Notation such as 173210±510 ± 710 denotes two types of measurement uncertainty. In the case of the top quark, the first uncertainty is statistical in nature, and the second is systematic.

INTERACTING QUARKS

As described by quantum chromodynamics, the strong interaction between quarks is mediated by gluons, massless vector gauge bosons. Each gluon carries one color charge and one anticolor charge. In the standard framework of particle interactions (part of a more general formulation known as perturbation theory), gluons are constantly exchanged between quarks through a virtual emission and absorption process. When a gluon is transferred between quarks, a color change occurs in both; for example, if a red quark emits a red–antigreen gluon, it becomes green, and if a green quark absorbs a red–antigreen gluon, it becomes red. Therefore, while each quark's color constantly changes, their strong interaction is preserved.

Since gluons carry color charge, they themselves are able to emit and absorb other gluons. This causes asymptotic freedom: as quarks come closer to each other, the chromodynamic binding force between them weakens.Conversely, as the distance between quarks increases, the binding force strengthens. The color field becomes stressed, much as an elastic band is stressed when stretched, and more gluons of appropriate color are spontaneously created to strengthen the field. Above a certain energy threshold, pairs of quarks and antiquarks are created. These pairs bind with the quarks being separated, causing new hadrons to form. This phenomenon is known as color confinement: quarks never appear in isolation. This process of hadronization occurs before quarks, formed in a high energy collision, are able to interact in any other way. The only exception is the top quark, which may decay before it hadronizes.

Sea Quarks

Hadrons contain, along with the valence quarks (qv) that contribute to their quantum numbers, virtual quark–antiquark (qq) pairs known as sea quarks (qs). Sea quarks form when a gluon of the hadron's color field splits; this process also works in reverse in that the annihilation of two sea quarks produces a gluon. The result is a constant flux of gluon splits and creations colloquially known as "the sea". Sea quarks are much less stable than their valence counterparts, and they typically annihilate each other within the interior of the hadron. Despite this, sea quarks can hadronize into baryonic or mesonic particles under certain circumstances.

Other Phases of Quark Matter

Under sufficiently extreme conditions, quarks may become deconfined and exist as free particles. In the course of asymptotic freedom, the strong interaction becomes weaker at higher temperatures. Eventually, color confinement would be lost and an extremely hot plasma of freely moving quarks and gluons would be formed. This theoretical phase of matter is called quark–gluon plasma. The exact conditions needed to give rise to this state are unknown and have been the subject of a great deal of speculation and experimentation. A recent estimate puts the needed temperature at $(1.90\pm0.02)\times10^{12}$ kelvin. While a state of entirely free quarks and gluons has never been achieved (despite numerous attempts by CERN in the 1980s and 1990s), recent experiments at the Relativistic Heavy Ion Collider have yielded evidence for liquid-like quark matter exhibiting "nearly perfect" fluid motion.

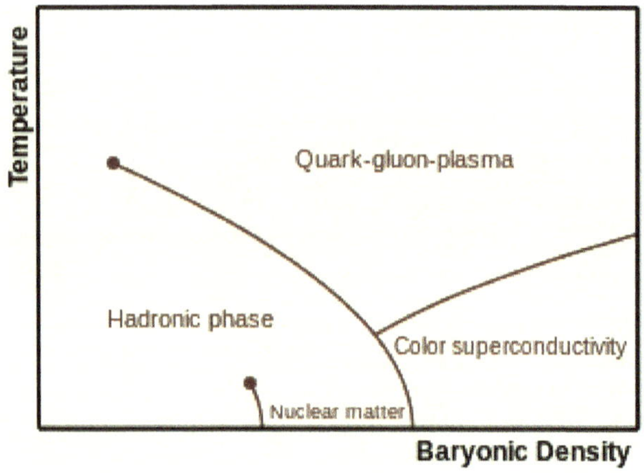

A qualitative rendering of the phase diagram of quark matter. The precise details of the diagram are the subject of ongoing research.

The quark–gluon plasma would be characterized by a great increase in the number of heavier quark pairs in relation to the number of up and down quark pairs. It is believed that in the period prior to 10^{-6} seconds after the Big Bang (the quark epoch), the universe was filled with quark–gluon plasma, as the temperature was too high for hadrons to be stable.

Given sufficiently high baryon densities and relatively low temperatures – possibly comparable to those found in neutron stars – quark matter is expected to degenerate into a Fermi liquid of weakly interacting quarks. This liquid would be characterized by a condensation of colored quark Cooper pairs, thereby breaking the local $SU(3)_c$ symmetry. Because quark Cooper pairs harbor color charge, such a phase of quark matter would be color superconductive; that is, color charge would be able to pass through it with no resistance.

Up Quark

The up quark or u quark (symbol: u) is the lightest of all quarks, a type of elementary particle, and a major constituent of matter. It, along with the down quark, forms the neutrons (one up quark, two down quarks) and protons (two up quarks, one down quark) of atomic nuclei. It is part of the first generation of matter, has an electric charge of +2/3 e and a bare mass of 1.8–3.0 MeV/c^2. Like all quarks, the up quark is an elementary fermion with spin 1/2, and experiences all four fundamental interactions: gravitation, electromagnetism, weak interactions, and strong interactions. The antiparticle of the up quark is the up antiquark (sometimes called antiup quark or simply antiup), which differs from it only in that some of its properties have equal magnitude but opposite sign.

Its existence (along with that of the down and strange quarks) was postulated in 1964 by Murray Gell-Mann and George Zweig to explain the Eightfold Way classification scheme of hadrons. The up quark was first observed by experiments at the Stanford Linear Accelerator Center in 1968.

History

In the beginnings of particle physics (first half of the 20th century), hadrons such as protons, neutrons and pions were thought to be elementary particles. However, as new hadrons were discovered, the 'particle zoo' grew from a few particles in the early 1930s and 1940s to several dozens of them in the 1950s. The relationships between each of them were unclear until 1961, when Murray Gell-Mann and Yuval Ne'eman (independently of each other) proposed a hadron classification scheme called the Eightfold Way, or in more technical terms, SU(3) flavor symmetry.

This classification scheme organized the hadrons into isospin multiplets, but the physical basis behind it was still unclear. In 1964, Gell-Mann and George Zweig (independently of each other) proposed the quark model, then consisting only of up, down, and strange quarks. However, while the quark model explained the Eightfold Way, no direct evidence of the existence of quarks was found until 1968 at the Stanford Linear Accelerator Center. Deep inelastic scattering experiments indicated that protons had substructure, and that protons made of three more-fundamental particles explained the data (thus confirming the quark model).

At first people were reluctant to describe the three bodies as quarks, instead preferring Richard Feynman's parton description, but over time the quark theory became accepted.

Mass

Despite being extremely common, the bare mass of the up quark is not well determined, but probably lies between 1.8 and 3.0 MeV/c^2. Lattice QCD calculations give a more precise value: 2.01±0.14 MeV/c^2.

When found in mesons (particles made of one quark and one antiquark) or baryons (particles made of three quarks), the 'effective mass' (or 'dressed' mass) of quarks becomes greater because of the binding energy caused by the gluon field between each quark. The bare mass of up quarks is so light, it cannot be straightforwardly calculated because relativistic effects have to be taken into account.

Down Quark

The down quark or d quark (symbol: d) is the second-lightest of all quarks, a type of elementary particle, and a major constituent of matter. Together with the up quark, it forms the neutrons (one up quark, two down quarks) and protons (two up quarks, one down quark) of atomic nuclei. It is part of the first generation of matter, has an electric charge of −1/3 e and a bare mass of 4.8+0.5−0.3 MeV/c^2. Like all quarks, the down quark is an elementary fermion with spin 1/2, and experiences all four fundamental interactions: gravitation, electromagnetism, weak interactions, and strong interactions. The antiparticle of the down quark is the down antiquark (sometimes called antidown quark or simply antidown), which differs from it only in that some of its properties have equal magnitude but opposite sign.

Its existence (along with that of the up and strange quarks) was postulated in 1964 by Murray Gell-Mann and George Zweig to explain the Eightfold Way classification scheme of hadrons. The down quark was first observed by experiments at the Stanford Linear Accelerator Center in 1968.

History

In the beginnings of particle physics (first half of the 20th century), hadrons such as protons, neutrons, and pions were thought to be elementary particles. However, as new hadrons were discovered, the 'particle zoo' grew from a few particles in the early 1930s and 1940s to several dozens of them in the 1950s. The relationships between each of them was unclear until 1961, when Murray Gell-Mann and Yuval Ne'eman (independently of each other) proposed a hadron classification scheme called the Eightfold Way, or in more technical terms, SU(3) flavor symmetry.

This classification scheme organized the hadrons into isospin multiplets, but the physical basis behind it was still unclear. In 1964, Gell-Mann and George Zweig (independently of each other) proposed the quark model, then consisting only of up, down, and strange quarks. However, while the quark model explained the Eightfold Way, no direct evidence of the existence of quarks was found until 1968 at the Stanford Linear Accelerator Center. Deep inelastic scattering experiments indicated that protons had substructure, and that protons made of three more-fundamental particles explained the data (thus confirming the quark model).

At first people were reluctant to identify the three-bodies as quarks, instead preferring Richard Feynman's parton description, but over time the quark theory became accepted.

Mass

Despite being extremely common, the bare mass of the down quark is not well determined, but probably lies between 4.5 and 5.3MeV/c^2. Lattice QCD calculations give a more precise value: 4.79±0.16 MeV/c^2.

When found in mesons (particles made of one quark and one antiquark) or baryons (particles made of three quarks), the 'effective mass' (or 'dressed' mass) of quarks becomes greater because of the binding energy caused by the gluon field between quarks. For example, the effective mass of down quarks in a proton is around 330 MeV/c^2. Because the bare mass of down quarks is so small, it cannot be straightforwardly calculated because relativistic effects have to be taken into account.

Charm Quark

The charm quark or c quark (from its symbol, c) is the third most massive of all quarks, a type of elementary particle. Charm quarks are found in hadrons, which are subatomic particles made of quarks. Example of hadrons containing charm quarks include the J/ψ meson (J/ψ), D mesons (D), charmed Sigma baryons (Σc), and other charmed particles.

It, along with the strange quark is part of the second generation of matter, and has an electric charge of +2/3 e and a bare mass of 1.29+0.05−0.11 GeV/c^2. Like all quarks, the charm quark is an elementary fermion with spin 1/2, and experiences all four fundamental interactions: gravitation, electromagnetism, weak interactions, and strong interactions.

The antiparticle of the charm quark is the charm antiquark (sometimes called anticharm quark or simply anticharm), which differs from it only in that some of its properties have equal magnitude but opposite sign.

The existence of a fourth quark had been speculated by a number of authors around 1964 (for instance by James Bjorken and Sheldon Glashow), but its prediction is usually credited to Sheldon Glashow, John Iliopoulos and Luciano Maiani in 1970. The first charmed particle (a particle containing a charm quark) to be discovered was the J/ψ meson. It was discovered by a team at the Stanford Linear Accelerator Center (SLAC), led by Burton Richter, and one at the Brookhaven National Laboratory (BNL), led by Samuel Ting.

The 1974 discovery of the J/ψ (and thus the charm quark) ushered in a series of breakthroughs which are collectively known as the November Revolution.

Hadrons Containing Charm Quarks

List of baryons and list of mesons

Some of the hadrons containing charm quarks include:

- D mesons contain a charm quark (or its antiparticle) and an up or down quark.
- Ds mesons contain a charm quark and a strange quark.
- There are many charmonium states, for example the J/ψ particle. These consist of a charm quark and its antiparticle.
- Charmed baryons have been observed, and are named in analogy with strange baryons (e.g. Λ+c).

Strange Quark

The strange quark or s quark (from its symbol, s) is the third-lightest of all quarks, a type of elementary particle. Strange quarks are found in subatomic particles called hadrons. Example of hadrons containing strange quarks include kaons (K), strange D mesons (Ds), Sigma baryons (Σ), and other strange particles.

Along with the charm quark, it is part of the second generation of matter, and has an electric charge of −1/3 e and a bare mass of 95+5−5 MeV/c^2. Like all quarks, the strange quark is an elementary fermion with spin 1/2, and experiences all four fundamental interactions: gravitation, electromagnetism, weak interactions, and strong interactions. The antiparticle of the strange quark is the strange antiquark (sometimes called antistrange quark or simply antistrange), which differs from it only in that some of its properties have equal magnitude but opposite sign.

The first strange particle (a particle containing a strange quark) was discovered in 1947 (kaons), but the existence of the strange quark itself (and that of the up and down quarks) was only postulated in 1964 by Murray Gell-Mann and George Zweig to explain the Eightfold Way classification scheme of hadrons. The first evidence for the existence of quarks came in 1968, in deep inelastic scattering experiments at the Stanford Linear

Accelerator Center. These experiments confirmed the existence of up and down quarks, and by extension, strange quarks, as they were required to explain the Eightfold Way.

Top Quark

The top quark, also known as the t quark (symbol: t) or truth quark, is the most massive of all observed elementary particles. Like all quarks, the top quark is an elementary fermion with spin1/2, and experiences all four fundamental interactions: gravitation, electromagnetism, weak interactions, and strong interactions. It has an electric charge of +2/3 e, It has a large mass of 172.44 ± 0.13 (stat) ± 0.47 (syst)GeV/c^2, which is about the same mass as an atom of tungsten. The antiparticle of the top quark is the top antiquark (symbol: t, sometimes called antitop quark or simply antitop), which differs from it only in that some of its properties have equal magnitude but opposite sign.

The top quark interacts primarily by the strong interaction, but can only decay through the weak force. It decays to a W boson and either a bottom quark (most frequently), a strange quark, or, on the rarest of occasions, a down quark. The Standard Model predicts its mean lifetime to be roughly 5×10^{-25} s. This is about a twentieth of the timescale for strong interactions, and therefore it does not form hadrons, giving physicists a unique opportunity to study a "bare" quark (all other quarks hadronize, meaning that they combine with other quarks to form hadrons, and can only be observed as such). Because it is so massive, the properties of the top quark allow predictions to be made of the mass of the Higgs boson under certain extensions of the Standard Model. As such, it is extensively studied as a means to discriminate between competing theories.

Its existence (and that of the bottom quark) was postulated in 1973 by Makoto Kobayashi and Toshihide Maskawa to explain the observed CP violations in kaon decay, and was discovered in 1995 by the CDF and DØ experiments at Fermilab. Kobayashi and Maskawa won the 2008 Nobel Prize in Physics for the prediction of the top and bottom quark, which together form the third generation of quarks.

Properties

- At the final Tevatron energy of 1.96 TeV, top–antitop pairs were produced with a cross section of about 7 picobarns (pb). The Standard Model prediction (at next-to-leading order with m_t = 175 GeV/c^2) is 6.7–7.5 pb.
- The W bosons from top quark decays carry polarization from the parent particle, hence pose themselves as a unique probe to top polarization.
- In the Standard Model, the top quark is predicted to have a spin quantum number of $\frac{1}{2}$ and electric charge $+\frac{2}{3}$. A first measurement of the top quark charge has been published, resulting in approximately 90% confidence limit that the top quark charge is indeed $+\frac{2}{3}$.

Production

Because top quarks are very massive, large amounts of energy are needed to create one. The only way to achieve such high energies is through high energy collisions. These occur naturally in the Earth's upper atmosphere as cosmic rays collide with particles in the air, or can be created in a particle accelerator. In 2011, after the Tevatron ceased operations, the Large Hadron Collider at CERN became the only accelerator that generates a beam of sufficient energy to produce top quarks, with a center-of-mass energy of 7 TeV. There are multiple processes that can lead to the production of top quarks, but they can be conceptually divided in two categories.

Top-Quark Pairs

The most common is production of a top–antitop pair via strong interactions. In a collision, a highly energetic gluon is created, which subsequently decays into a top and antitop. This process was responsible for the majority of the top events at Tevatron and was the process observed when the top was first discovered in 1995. It is also possible to produce pairs of top–antitop through the decay of an intermediate photon or Z-boson. However, these processes are predicted to be much rarer and have a virtually identical experimental signature in a hadron collider like Tevatron.

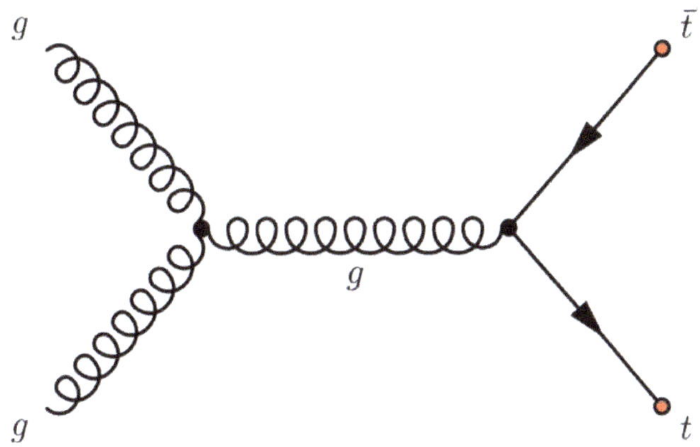

gluon-gluon fusion

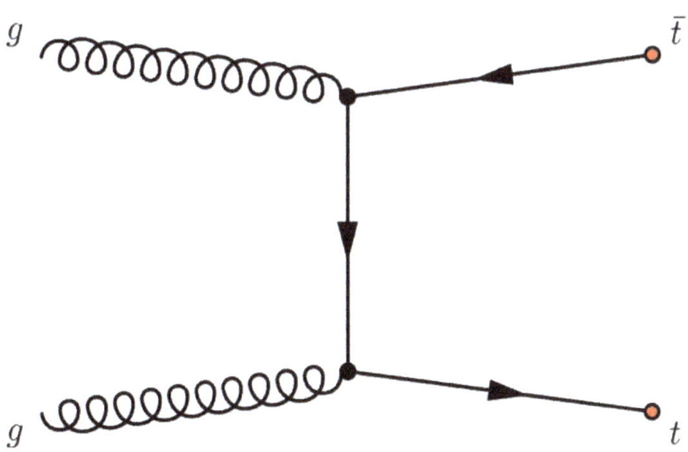

t-channel

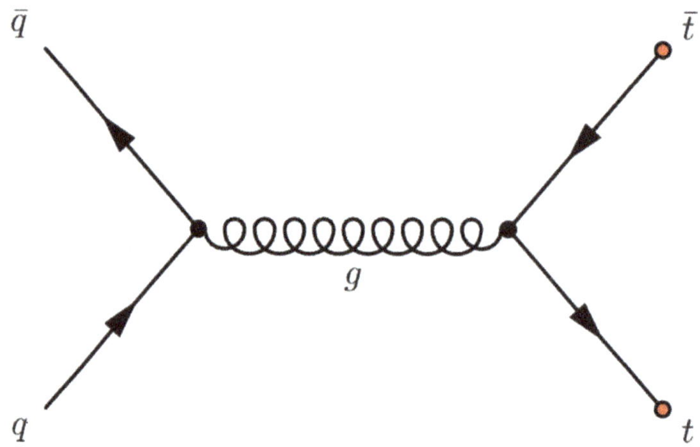

Quark-antiquark annihilation.

Single Top Quarks

A distinctly different process is the production of single top quarks via weak interaction. This can happen in several ways (called channels): either an intermediate W-boson decays into a top and antibottom quark ("s-channel") or a bottom quark (probably created in a pair through the decay of a gluon) transforms to a top quark by exchanging a W-boson with an up or down quark ("t-channel").

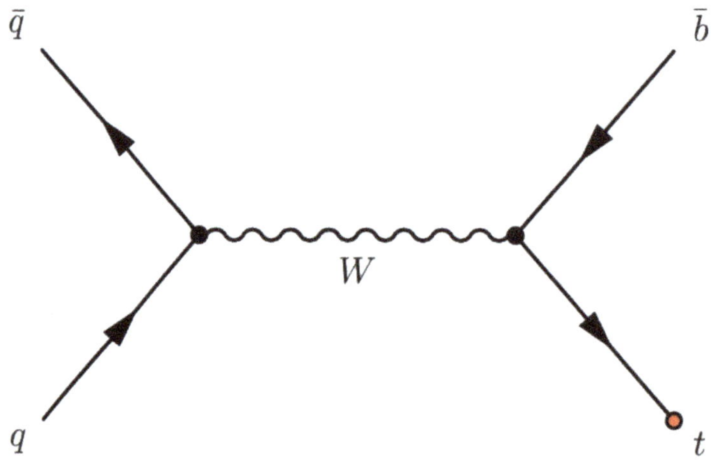

s-channel

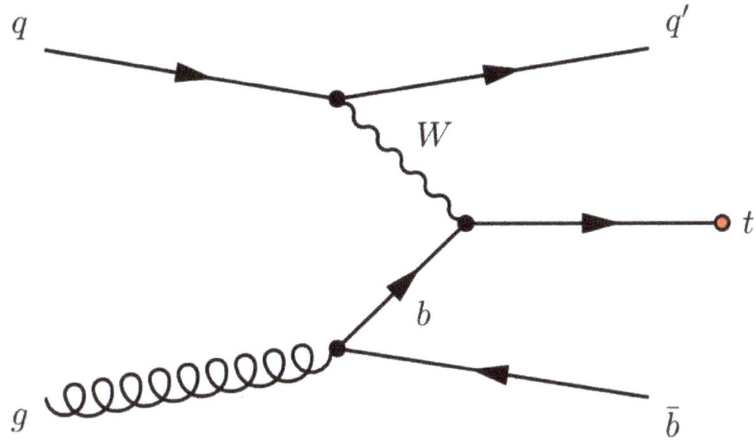

t-channel

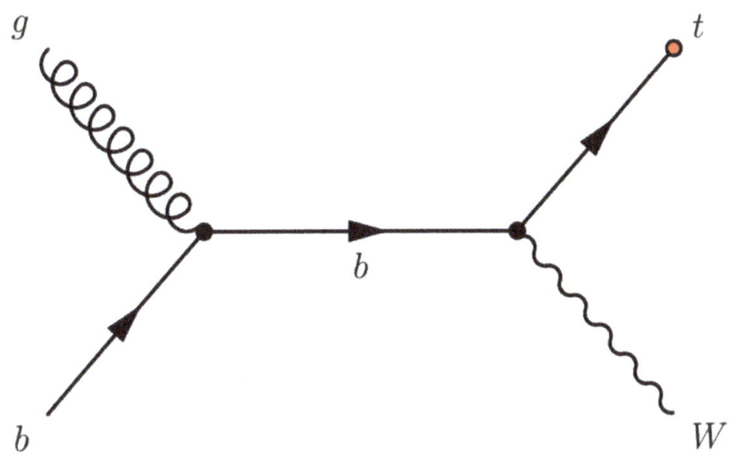

tW channel

A single top quark can also be produced in association with a W boson, requiring an initial state bottom quark ("tW-channel"). The first evidence for these processes was published by the DØ collaboration in December 2006, and in March 2009 the CDF and DØ collaborations released twin papers with the definitive observation of these processes. The main significance of measuring these production processes is that their frequency is directly proportional to the $|V_{tb}|^2$ component of the CKM matrix.

Decay

Because of its enormous mass, the top quark is extremely short-lived with a predicted lifetime of only 5×10^{-25} s. As a result, top quarks do not have time to form hadrons before they decay, as other quarks do which provides physicists with the unique opportunity to study the behavior of a "bare" quark. The only known way the top quark can decay is through the weak interaction producing a W-boson and a down-type quark (down, strange, or bottom).

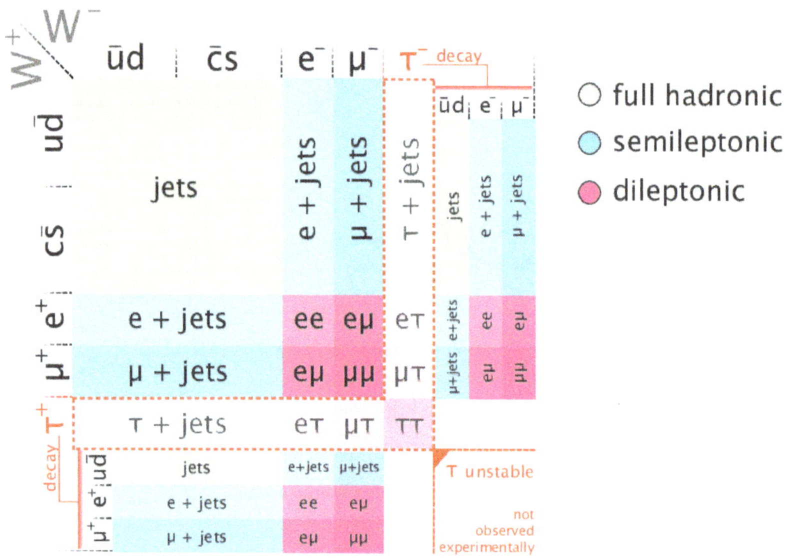

All possible final states of the decay of a top-quark pair.

In particular, it is possible to directly determine the branching ratio $\Gamma(W^{+}b) / \Gamma(W^{+}q$ $(q = b,s,d))$. The best current determination of this ratio is 0.91±0.04. Since this ratio is equal to $|V_{tb}|^2$ according to the Standard Model, this gives another way of determining the CKM element $|V_{tb}|$, or in combination with the determination of $|V_{tb}|$ from single top production provides tests for the assumption that the CKM matrix is unitary.

The Standard Model also allows more exotic decays, but only at one loop level, meaning that they are extremely suppressed. In particular, it is possible for a top quark to decay into another up-type quark (an up or a charm) by emitting a photon or a Z-boson. Searches for these exotic decay modes have provided no evidence for their existence in accordance with expectations from the Standard Model. The branching ratios for these decays have been determined to be less than 5.9 in 1,000 for photonic decay and less than 2.1 in 1,000 for Z-boson decay at 95% confidence.

Mass and Coupling to the Higgs Boson

The Standard Model describes fermion masses through the Higgs mechanism. The Higgs boson has a Yukawa coupling to the left- and right-handed top quarks. After electroweak symmetry breaking (when the Higgs acquires a vacuum expectation value), the left- and right-handed components mix, becoming a mass term.

$$\mathcal{L} = y_t h q u^c \rightarrow \frac{y_t v}{\sqrt{2}}(1 + h^0/v)uu^c$$

The top quark Yukawa coupling has a value of

$$y_t = \sqrt{2}m_t/v \simeq 1$$

where v = 246 GeV is the value of the Higgs vacuum expectation value.

Yukawa Couplings

In the Standard Model, all of the quark and lepton Yukawa couplings are small compared to the top quark Yukawa coupling. Understanding this hierarchy in the fermion masses is an open problem in theoretical physics. Yukawa couplings are not constants and their values change depending on the energy scale (distance scale) at which they are measured. The dynamics of Yukawa couplings are determined by the renormalization group equation.

One of the prevailing views in particle physics is that the size of the top quark Yukawa coupling is determined by the renormalization group, leading to the "quasi-infrared fixed point."

The Yukawa couplings of the up, down, charm, strange and bottom quarks, are hypothesized to have small values at the extremely high energy scale of grand unification, 10^{15} GeV. They increase in value at lower energy scales, at which the quark masses are generated by the Higgs. The slight growth is due to corrections from the QCD coupling. The corrections from the Yukawa couplings are negligible for the lower mass quarks.

If, however, a quark Yukawa coupling has a large value at very high energies, its Yukawa corrections will evolve and cancel against the QCD corrections. This is known as a (quasi-) infrared fixed point. No matter what the initial starting value of the coupling is, if it is sufficiently large it will reach this fixed point value. The corresponding quark mass is then predicted.

The top quark Yukawa coupling lies very near the infrared fixed point of the Standard Model. The renormalization group equation is:

$$\mu\frac{\partial}{\partial\mu}y_t \approx \frac{y_t}{16\pi^2}\left(\frac{9}{2}y_t^2 - 8g_3^2 - \frac{9}{4}g_2^2 - \frac{17}{20}g_1^2\right),$$

where g_3 is the color gauge coupling, g_2 is the weak isospin gauge coupling, and g_1 is the weak hypercharge gauge coupling. This equation describes how the Yukawa coupling changes with energy scale μ. Solutions to this equation for large initial values y_t cause the right-hand side of the equation to quickly approach zero, locking y_t to the QCD coupling g_3. The value of the fixed point is fairly precisely determined in the Standard Model, leading to a top quark mass of 230 GeV. However, if there is more than one Higgs doublet, the mass value will be reduced by Higgs mixing angle effects in an unpredicted way.

In the minimal supersymmetric extension of the Standard Model (MSSM), there are two Higgs doublets and the renormalization group equation for the top quark Yukawa coupling is slightly modified:

$$\mu\frac{\partial}{\partial\mu}y_t \approx \frac{y_t}{16\pi^2}\left(6y_t^2 + y_b^2 - \frac{16}{3}g_3^2 - 3g_2^2 - \frac{13}{15}g_1^2\right),$$

where y_b is the bottom quark Yukawa coupling. This leads to a fixed point where the top mass is smaller, 170–200 GeV. The uncertainty in this prediction arises because the bottom quark Yukawa coupling can be amplified in the MSSM. Some theorists believe this is supporting evidence for the MSSM.

The quasi-infrared fixed point has subsequently formed the basis of top quark condensation theories of electroweak symmetry breaking in which the Higgs boson is composite at extremely short distance scales, composed of a pair of top and antitop quarks.

Bottom Quark

The bottom quark or b quark, also known as the beauty quark, is a third-generation quark with a charge of −1/3 e. Although all quarks are described in a similar way by quantum chromodynamics, the bottom quark's large bare mass (around 4.2 GeV/c^2, a bit more than four times the mass of a proton), combined with low values of the CKM matrix elements V_{ub} and V_{cb}, gives it a distinctive signature that makes it relatively easy to identify experimentally (using a technique called B-tagging). Because three generations of quark are required for CP violation, mesons containing the bottom quark are the easiest particles to use to investigate the phenomenon; such experiments are being performed at the BaBar, Belle and LHCb experiments.

The bottom quark is also notable because it is a product in almost all top quark decays, and is a frequent decay product for the Higgs boson. The bottom quark was theorized in 1973 by physicists Makoto Kobayashi and Toshihide Maskawa to explain CP violation. The name "bottom" was introduced in 1975 by Haim Harari. The bottom quark was discovered in 1977 by the Fermilab E288 experiment team led by Leon M. Lederman, when collisions produced bottomonium. Kobayashi and Maskawa won the 2008 Nobel Prize in Physics for their explanation of CP-violation. On its discovery, there were efforts to name the bottom quark "beauty", but "bottom" became the predominant usage.

The bottom quark can decay into either an up quark or charm quark via the weak interaction. Both these decays are suppressed by the CKM matrix, making lifetimes of most bottom particles (~10^{-12} s) somewhat higher than those of charmed particles (~10^{-13} s), but lower than those of strange particles (from ~10^{-10} to ~10^{-8} s).

REFERENCES

1. Notes on Dirac's lecture Developments in Atomic Theory at Le Palais de la Découverte, 6 December 1945, UKNATARCHI Dirac Papers BW83/2/257889. See note 64 on page 331 in "The Strangest Man: The Hidden Life of Paul Dirac, Mystic of the Atom" by Graham Farmelo
2. Physical Review D volume 87, page 0550003, year 2013, author Weiner, Richard M., title "Spin-statistics-quantum number connection and supersymmetry" arxiv:1302.0969
3. T. Morii; C. S. Lim; S. N. Mukherjee (1 January 2004). The Physics of the Standard Model and Beyond. World Scientific. ISBN 978-981-279-560-1.
4. Weiner, Richard M. (2010). "The Mysteries of Fermions". International Journal

of Theoretical Physics. 49 (5): 1174–1180. arXiv:0901.3816Freely accessible. Bibcode:2010IJTP...49.1174W. doi:10.1007/s10773-010-0292-7.

5. "Quark (subatomic particle)". Encyclopædia Britannica. Retrieved 2008-06-29.
6. R. Nave. "Confinement of Quarks". HyperPhysics. Georgia State University, Department of Physics and Astronomy. Retrieved 2008-06-29.
7. R. Nave. "Bag Model of Quark Confinement". HyperPhysics. Georgia State University, Department of Physics and Astronomy. Retrieved 2008-06-29.
8. R. Nave. "Quarks". HyperPhysics. Georgia State University, Department of Physics and Astronomy. Retrieved 2008-06-29.
9. B. Carithers; P. Grannis (1995). "Discovery of the Top Quark" (PDF). Beam Line. SLAC. 25 (3): 4–16. Retrieved 2008-09-23.
10. E.D. Bloom; et al. (1969). "High-Energy Inelastic e–p Scattering at 6° and 10°". Physical Review Letters. 23 (16): 930–934. Bibcode:1969PhRvL..23..930B. doi:10.1103/PhysRevLett.23.930.
11. M. Breidenbach; et al. (1969). "Observed Behavior of Highly Inelastic Electron–Proton Scattering". Physical Review Letters. 23 (16): 935–939. Bibcode:1969PhRvL..23..935B. doi:10.1103/PhysRevLett.23.935.
12. S.S.M. Wong (1998). Introductory Nuclear Physics (2nd ed.). Wiley Interscience. p. 30. ISBN 0-471-23973-9.
13. K.A. Peacock (2008). The Quantum Revolution. Greenwood Publishing Group. p. 125. ISBN 0-313-33448-X.
14. B. Povh; C. Scholz; K. Rith; F. Zetsche (2008). Particles and Nuclei. Springer. p. 98. ISBN 3-540-79367-4.
15. Section 6.1. in P.C.W. Davies (1979). The Forces of Nature. Cambridge University Press. ISBN 0-521-22523-X.
16. M. Munowitz (2005). Knowing. Oxford University Press. p. 35. ISBN 0-19-516737-6.
17. W.-M. Yao; et al. (Particle Data Group) (2006). "Review of Particle Physics: Pentaquark Update" (PDF). Journal of Physics G. 33 (1): 1–1232. arXiv:astro-ph/0601168Freely accessible. Bibcode:2006JPhG...33....1Y. doi:10.1088/0954-3899/33/1/001Freely accessible.
18. C. Amsler; et al. (Particle Data Group) (2008). "Review of Particle Physics: Pentaquarks" (PDF). Physics Letters B. 667 (1): 1–1340. Bibcode:2008PhLB..667....1A. doi:10.1016/j.physletb.2008.07.018.
19. C. Amsler; et al. (Particle Data Group) (2008). "Review of Particle Physics: New Charmonium-Like States" (PDF). Physics Letters B. 667 (1): 1–1340. Bibcode:2008PhLB..667....1A. doi:10.1016/j.physletb.2008.07.018.
20. E.V. Shuryak (2004). The QCD Vacuum, Hadrons and Superdense Matter. World Scientific. p. 59. ISBN 981-238-574-6.
21. R. Aaij; et al. (LHCb collaboration) (2015). "Observation of J/ψp resonances consistent with pentaquark states in Λ0

22. b→J/ψK−p decays". Physical Review Letters. 115 (7): 072001. arXiv:1507.03414Freely accessible. Bibcode:2015PhRvL.115g2001A. doi:10.1103/PhysRevLett.115.072001Freely accessible.
23. C. Amsler; et al. (Particle Data Group) (2008). "Review of Particle Physics: b′ (4th Generation) Quarks, Searches for" (PDF). Physics Letters B. 667 (1): 1–1340. Bibcode:2008PhLB..667....1A. doi:10.1016/j.physletb.2008.07.018.
24. C. Amsler; et al. (Particle Data Group) (2008). "Review of Particle Physics: t′ (4th Generation) Quarks, Searches for" (PDF). Physics Letters B. 667 (1): 1–1340. Bibcode:2008PhLB..667....1A. doi:10.1016/j.physletb.2008.07.018.
25. D. Decamp; et al. (ALEPH Collaboration) (1989). "Determination of the number of light neutrino species". Physics Letters B. 231 (4): 519. Bibcode:1989PhLB..231..519D. doi:10.1016/0370-2693(89)90704-1.
26. A.Fisher (1991). "Searching for the Beginning of Time: Cosmic Connection". Popular Science. 238 (4): 70.
27. J.D. Barrow (1997) [1994]. "The Singularity and Other Problems". The Origin of the Universe (Reprint ed.). Basic Books. ISBN 978-0-465-05314-8.
28. D.H. Perkins (2003). Particle Astrophysics. Oxford University Press. p. 4. ISBN 0-19-850952-9.
29. M. Gell-Mann (1964). "A Schematic Model of Baryons and Mesons". Physics Letters. 8 (3): 214–215. Bibcode:1964PhL.....8..214G. doi:10.1016/S0031-9163(64)92001-3.
30. G. Zweig (1964). "An SU(3) Model for Strong Interaction Symmetry and its Breaking" (PDF). CERN Report No.8182/TH.401.
31. G. Zweig (1964). "An SU(3) Model for Strong Interaction Symmetry and its Breaking: II". CERN Report No.8419/TH.412.
32. M. Gell-Mann (2000) [1964]. "The Eightfold Way: A theory of strong interaction symmetry". In M. Gell-Mann, Y. Ne›eman. The Eightfold Way. Westview Press. p. 11. ISBN 0-7382-0299-1.
33. Original: M. Gell-Mann (1961). "The Eightfold Way: A theory of strong interaction symmetry". Synchrotron Laboratory Report CTSL-20. California Institute of Technology.
34. Y. Ne›eman (2000) [1964]. "Derivation of strong interactions from gauge invariance". In M. Gell-Mann, Y. Ne›eman. The Eightfold Way. Westview Press. ISBN 0-7382-0299-1.
35. Original Y. Ne›eman (1961). "Derivation of strong interactions from gauge invariance". Nuclear Physics. 26 (2): 222. Bibcode:1961NucPh..26..222N. doi:10.1016/0029-5582(61)90134-1.
36. R.C. Olby; G.N. Cantor (1996). Companion to the History of Modern Science. Taylor & Francis. p. 673. ISBN 0-415-14578-3.
37. A.Pickering (1984). Constructing Quarks. University of Chicago Press. pp. 114–125. ISBN 0-226-66799-5.
38. B.J. Bjorken; S.L. Glashow (1964). "Elementary Particles and SU(4)". Physics

Letters. 11 (3): 255–257. Bibcode:1964PhL....11..255B. doi:10.1016/0031-9163(64)90433-0.

39. J.I. Friedman. "The Road to the Nobel Prize". Hue University. Retrieved 2008-09-29.
40. R.P. Feynman (1969). "Very High-Energy Collisions of Hadrons". Physical Review Letters. 23 (24): 1415–1417. Bibcode:1969PhRvL..23.1415F. doi:10.1103/PhysRevLett.23.1415.
41. S. Kretzer; H.L. Lai; F.I. Olness; W.K. Tung (2004). "CTEQ6 Parton Distributions with Heavy Quark Mass Effects". Physical Review D. 69 (11): 114005. arXiv:hep-ph/0307022Freely accessible. Bibcode:2004PhRvD..69k4005K. doi:10.1103/PhysRevD.69.114005.
42. D.J. Griffiths (1987). Introduction to Elementary Particles. John Wiley & Sons. p. 42. ISBN 0-471-60386-4.
43. M.E. Peskin; D.V. Schroeder (1995). An introduction to quantum field theory. Addison–Wesley. p. 556. ISBN 0-201-50397-2.
44. V.V. Ezhela (1996). Particle physics. Springer. p. 2. ISBN 1-56396-642-5.
45. S.L. Glashow; J. Iliopoulos; L. Maiani (1970). "Weak Interactions with Lepton–Hadron Symmetry". Physical Review D. 2 (7): 1285–1292. Bibcode:1970PhRvD...2.1285G. doi:10.1103/PhysRevD.2.1285.
46. D.J. Griffiths (1987). Introduction to Elementary Particles. John Wiley & Sons. p. 44. ISBN 0-471-60386-4.
47. M. Kobayashi; T. Maskawa (1973). "CP-Violation in the Renormalizable Theory of Weak Interaction". Progress of Theoretical Physics. 49 (2): 652–657. Bibcode:1973PThPh..49..652K. doi:10.1143/PTP.49.652Freely accessible.
48. H. Harari (1975). "A new quark model for hadrons". Physics Letters B. 57 (3): 265. Bibcode:1975PhLB...57..265H. doi:10.1016/0370-2693(75)90072-6.
49. K.W. Staley (2004). The Evidence for the Top Quark. Cambridge University Press. pp. 31–33. ISBN 978-0-521-82710-2.
50. S.W. Herb; et al. (1977). "Observation of a Dimuon Resonance at 9.5 GeV in 400-GeV Proton-Nucleus
51. M. Bartusiak (1994). A Positron named Priscilla. National Academies Press. p. 245. ISBN 0-309-04893-1.
52. Collisions at $\sqrt{s}$ = 1.8 TeV". Physical Review Letters. 74 (13): 2422–2426. Bibcode:1995PhRvL..74.2422A. doi:10.1103/PhysRevLett.74.2422.
53. K.W. Staley (2004). The Evidence for the Top Quark. Cambridge University Press. p. 144. ISBN 0-521-82710-8.
54. "New Precision Measurement of Top Quark Mass". Brookhaven National Laboratory News. 2004. Retrieved 2013-11-03.
55. J. Joyce (1982) [1939]. Finnegans Wake. Penguin Books. p. 383. ISBN 0-14-006286-6.
56. M. Gell-Mann (1995). The Quark and the Jaguar: Adventures in the Simple and

the Complex. Henry Holt and Co. p. 180. ISBN 978-0-8050-7253-2.

57. J. Gleick (1992). Genius: Richard Feynman and modern physics. Little Brown and Company. p. 390. ISBN 0-316-90316-7.
58. J.J. Sakurai (1994). S.F Tuan, ed. Modern Quantum Mechanics (Revised ed.). Addison–Wesley. p. 376. ISBN 0-201-53929-2.
59. D.H. Perkins (2000). Introduction to high energy physics. Cambridge University Press. p. 8. ISBN 0-521-62196-8.
60. M. Riordan (1987). The Hunting of the Quark: A True Story of Modern Physics. Simon & Schuster. p. 210. ISBN 978-0-671-50466-3.
61. F. Close (2006). The New Cosmic Onion. CRC Press. p. 133. ISBN 1-58488-798-2.
62. J.T. Volk; et al. (1987). "Letter of Intent for a Tevatron Beauty Factory" (PDF). Fermilab Proposal #783.
63. G. Fraser (2006). The New Physics for the Twenty-First Century. Cambridge University Press. p. 91. ISBN 0-521-81600-9.
64. "The Standard Model of Particle Physics". BBC. 2002. Retrieved 2009-04-19.
65. F. Close (2006). The New Cosmic Onion. CRC Press. pp. 80–90. ISBN 1-58488-798-2.
66. D. Lincoln (2004). Understanding the Universe. World Scientific. p. 116. ISBN 981-238-705-6.
67. "Weak Interactions". Virtual Visitor Center. Stanford Linear Accelerator Center. 2008. Retrieved 2008-09-28.
68. K. Nakamura; et al. (Particle Data Group) (2010). "Review of Particles Physics: The CKM Quark-Mixing Matrix" (PDF). Journal of Physics G. 37: 075021. Bibcode:2010JPhG...37g5021N. doi:10.1088/0954-3899/37/7A/075021Freely accessible.
69. Z. Maki; M. Nakagawa; S. Sakata (1962). "Remarks on the Unified Model of Elementary Particles". Progress of Theoretical Physics. 28 (5): 870. Bibcode:1962PThPh..28..870M. doi:10.1143/PTP.28.870Freely accessible.
70. B.C. Chauhan; M. Picariello; J. Pulido; E. Torrente-Lujan (2007). "Quark–lepton complementarity, neutrino and standard model data predict θPMNS
71. 13 = 9°+1°
72. −2°". European Physical Journal. C50 (3): 573–578. arXiv:hep-ph/0605032Freely accessible. Bibcode:2007EPJC...50..573C. doi:10.1140/epjc/s10052-007-0212-z.
73. R. Nave. "The Color Force". HyperPhysics. Georgia State University, Department of Physics and Astronomy. Retrieved 2009-04-26.
74. B.A. Schumm (2004). Deep Down Things. Johns Hopkins University Press. pp. 131–132. ISBN 0-8018-7971-X. OCLC 55229065.
75. Part III of M.E. Peskin; D.V. Schroeder (1995). An Introduction to Quantum Field Theory. Addison–Wesley. ISBN 0-201-50397-2.

76. V. Icke (1995). The force of symmetry. Cambridge University Press. p. 216. ISBN 0-521-45591-X.

77. M.Y. Han (2004). A story of light. World Scientific. p. 78. ISBN 981-256-034-3.

78. C. Sutton. "Quantum chromodynamics (physics)". Encyclopædia Britannica Online. Retrieved 2009-05-12.

79. A.Watson (2004). The Quantum Quark. Cambridge University Press. pp. 285–286. ISBN 0-521-82907-0.

80. K.A. Olive; et al. (Particle Data Group) (2014). "Review of Particle Physics". Chinese Physics C. 38 (9): 090001. Bibcode:2014ChPhC..38i0001O. doi:10.1088/1674-1137/38/9/090001Freely accessible.

81. W. Weise; A.M. Green (1984). Quarks and Nuclei. World Scientific. pp. 65–66. ISBN 9971-966-61-1.

82. D. McMahon (2008). Quantum Field Theory Demystified. McGraw–Hill. p. 17. ISBN 0-07-154382-1.

83. S.G. Roth (2007). Precision electroweak physics at electron–positron colliders. Springer. p. VI. ISBN 3-540-35164-7.

84. R.P. Feynman (1985). QED: The Strange Theory of Light and Matter (1st ed.). Princeton University Press. pp. 136–137. ISBN 0-691-08388-6.

85. M. Veltman (2003). Facts and Mysteries in Elementary Particle Physics. World Scientific. pp. 45–47. ISBN 981-238-149-X.

86. F. Wilczek; B. Devine (2006). Fantastic Realities. World Scientific. p. 85. ISBN 981-256-649-X.

87. F. Wilczek; B. Devine (2006). Fantastic Realities. World Scientific. pp. 400ff. ISBN 981-256-649-X.

88. M. Veltman (2003). Facts and Mysteries in Elementary Particle Physics. World Scientific. pp. 295–297. ISBN 981-238-149-X.

89. T. Yulsman (2002). Origin. CRC Press. p. 55. ISBN 0-7503-0765-X.

90. F. Garberson (2008). "Top Quark Mass and Cross Section Results from the Tevatron". arXiv:0808.0273Freely accessible [hep-ex].

91. J. Steinberger (2005). Learning about Particles. Springer. p. 130. ISBN 3-540-21329-5.

92. C.-Y. Wong (1994). Introduction to High-energy Heavy-ion Collisions. World Scientific. p. 149. ISBN 981-02-0263-6.

93. S.B. Rüester; V. Werth; M. Buballa; I.A. Shovkovy; D.H. Rischke (2005). "The phase diagram of neutral quark matter: Self-consistent treatment of quark masses". Physical Review D. 72 (3): 034003. arXiv:hep-ph/0503184Freely accessible. Bibcode:2005PhRvD..72c4004R. doi:10.1103/PhysRevD.72.034004.

94. M.G. Alford; K. Rajagopal; T. Schaefer; A. Schmitt (2008). "Color superconductivity in dense quark matter". Reviews of Modern Physics. 80 (4): 1455–1515. arXiv:0709.4635Freely accessible. Bibcode:2008RvMP...80.1455A. doi:10.1103/RevModPhys.80.1455.

95. S. Mrowczynski (1998). "Quark–Gluon Plasma". Acta Physica Polonica B. 29: 3711. arXiv:nucl-th/9905005Freely accessible. Bibcode:1998AcPPB..29.3711M Freely accessible.

96. Z. Fodor; S.D. Katz (2004). "Critical point of QCD at finite T and μ, lattice results for physical quark masses". Journal of High Energy Physics. 2004 (4): 50. arXiv:hep-lat/0402006Freely accessible. Bibcode:2004JHEP...04..050F. doi:10.1088/1126-6708/2004/04/050Freely accessible.

97. U. Heinz; M. Jacob (2000). "Evidence for a New State of Matter: An Assessment of the Results from the CERN Lead Beam Programme". arXiv:nucl-th/0002042Freely accessible.

98. "RHIC Scientists Serve Up "Perfect" Liquid". Brookhaven National Laboratory. 2005. Retrieved 2009-05-22.

99. T. Yulsman (2002). Origins: The Quest for Our Cosmic Roots. CRC Press. p. 75. ISBN 0-7503-0765-X.

100. Sedrakian; J.W. Clark; M.G. Alford (2007). Pairing in fermionic systems. World Scientific. pp. 2–3. ISBN 981-256-907-3.

CHAPTER 4

Bosons

Chapter Contents

- **Introduction**
- **Types**
- **Properties**
- **Gauge Boson**
- **Scalar Boson**
- **Higgs Boson**
- **Significance**

All fundamental particles, such as electrons and photons, have an intrinsic degree of freedom called "spin", which in many situations acts like a form of angular momentum. A boson is a particle with a total spin that is equal to an integer (0,1,2,...) times the Planck constant ($\hbar$). For example, photons (the particles that compose light) are bosons having total spins of just $\hbar$. This is in constrast to electrons, which are fermions and have spins of $\hbar/2$. All fundamental force carrying particles (photon, gluon, W, Z) are bosons.

INTRODUCTION

In quantum mechanics, a boson is a particle that follows Bose–Einstein statistics. Bosons make up one of the two classes of particles, the other being fermions. The name boson was coined by Paul Dirac to commemorate the contribution of the Indian physicist Satyendra Nath Bose in developing, with Einstein, Bose–Einstein statistics—which theorizes the characteristics of elementary particles. Examples of bosons include fundamental particles such as photons, gluons, and W and Z bosons (the four force-carrying gauge bosons of the Standard Model), the recently discovered Higgs boson, and the hypothetical graviton of quantum gravity; composite particles (e.g. mesons and stable nuclei of even mass number such as deuterium (with one proton and one neutron, mass number = 2), helium-4, or lead-208); and some quasiparticles (e.g. Cooper pairs, plasmons, and phonons).

Satyendra Nath Bose.

An important characteristic of bosons is that their statistics do not restrict the number of them that occupy the same quantum state. This property is exemplified by helium-4 when it is cooled to become a superfluid. Unlike bosons, two identical fermions cannot occupy the same quantum space. Whereas the elementary particles that make up matter (i.e. leptons and quarks) are fermions, the elementary bosons are force carriers that function as the 'glue' holding matter together. This property holds for all particles with integer spin (s = 0, 1, 2, etc.) as a consequence of the spin–statistics theorem. When a gas of Bose particles is cooled down to temperatures very close to absolute zero then the kinetic energy of the particles decreases to a negligible amount and they condense into a lowest energy level state. This state is called Bose-Einstein condensation. It is believed that this property is the explanation of superfluidity.

TYPES

Bosons may be either elementary, like photons, or composite, like mesons.

While most bosons are composite particles, in the Standard Model there are five bosons which are elementary:

- the four gauge bosons (γ · g · Z · W±)
- the one scalar boson (the Higgs boson (H0))

Additionally, the graviton (G) is a hypothetical elementary particle not incorporated in the Standard Model. If it exists, a graviton must be a boson, and could conceivably be a gauge boson.

Composite bosons are important in superfluidity and other applications of Bose–

Einstein condensates. When a gas of Bose particles is cooled to temperatures very close to absolute zero, then its kinetic energy decreases down to a negligible amount and the particles condense into the lowest energy state. This phenomenon is known as Bose-Einstein condensation and it is believed that this phenomenon is the secret behind superfluidity of liquids.

PROPERTIES

Bosons differ from fermions, which obey Fermi–Dirac statistics. Two or more identical fermions cannot occupy the same quantum state.

Since bosons with the same energy can occupy the same place in space, bosons are often force carrier particles. Fermions are usually associated with matter (although in quantum mechanics the distinction between the two concepts is not clear cut).

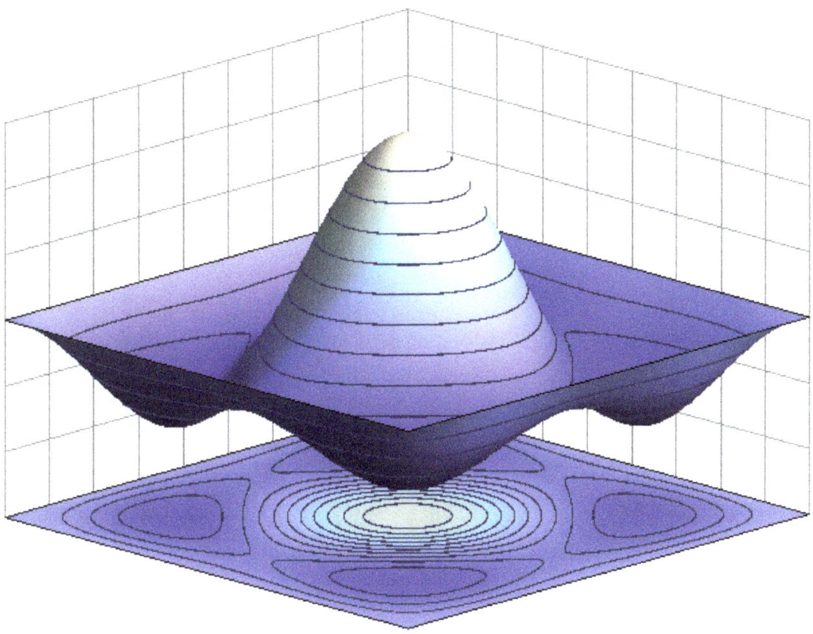

Symmetric wavefunction for a (bosonic) 2-particle state in an infinite square well potential.

Bosons are particles which obey Bose–Einstein statistics: when one swaps two bosons (of the same species), the wavefunction of the system is unchanged. Fermions, on the other hand, obey Fermi–Dirac statistics and the Pauli exclusion principle: two fermions cannot occupy the same quantum state, accounting for the “rigidity” or “stiffness” of matter which includes fermions. Thus fermions are sometimes said to be the constituents of matter, while bosons are said to be the particles that transmit interactions (force carriers), or the constituents of radiation. The quantum fields of bosons are bosonic fields, obeying canonical commutation relations.

The properties of lasers and masers, superfluid helium-4 and Bose–Einstein condensates are all consequences of statistics of bosons. Another result is that the spectrum of a photon gas in thermal equilibrium is a Planck spectrum, one example of which is black-body radiation; another is the thermal radiation of the opaque early Universe seen today as microwave background radiation. Interactions between elementary particles are called fundamental interactions. The fundamental interactions of virtual bosons with real particles result in all forces we know.

All known elementary and composite particles are bosons or fermions, depending on their spin: particles with half-integer spin are fermions; particles with integer spin are bosons. In the framework of nonrelativistic quantum mechanics, this is a purely empirical observation. However, in relativistic quantum field theory, the spin–statistics theorem shows that half-integer spin particles cannot be bosons and integer spin particles cannot be fermions.

In large systems, the difference between bosonic and fermionic statistics is only apparent at large densities—when their wave functions overlap. At low densities, both types of statistics are well approximated by Maxwell–Boltzmann statistics, which is described by classical mechanics.

Elementary Bosons

All observed elementary particles are either fermions or bosons. The observed elementary bosons are all gauge bosons: photons, W and Z bosons, gluons, except the Higgs boson which is a scalar boson.

- Photons are the force carriers of the electromagnetic field.
- W and Z bosons are the force carriers which mediate the weak force.
- Gluons are the fundamental force carriers underlying the strong force.
- Higgs bosons give W and Z bosons mass via the Higgs mechanism. Their existence was confirmed by CERN on 14 March 2013.

Finally, many approaches to quantum gravity postulate a force carrier for gravity, the graviton, which is a boson of spin plus or minus two.

Composite Bosons

Composite particles (such as hadrons, nuclei, and atoms) can be bosons or fermions depending on their constituents. More precisely, because of the relation between spin and statistics, a particle containing an even number of fermions is a boson, since it has integer spin.

Examples include the following:

- Any meson, since mesons contain one quark and one antiquark.
- The nucleus of a carbon-12 atom, which contains 6 protons and 6 neutrons.
- The helium-4 atom, consisting of 2 protons, 2 neutrons and 2 electrons.

The number of bosons within a composite particle made up of simple particles bound with a potential has no effect on whether it is a boson or a fermion.

To which States Can Bosons Crowd?

Bose–Einstein statistics encourages identical bosons to crowd into one quantum state, but not any state is necessarily convenient for it. Aside of statistics, bosons can interact – for example, helium-4 atoms are repulsed by intermolecular force on a very close approach, and if one hypothesizes their condensation in a spatially-localized state, then gains from the statistics cannot overcome a prohibitive force potential. A spatially-delocalized state (i.e. with low $|\psi(x)|$) is preferable: if the number density of the condensate is about the same as in ordinary liquid or solid state, then the repulsive potential for the N-particle condensate in such state can be no higher than for a liquid or a crystalline lattice of the same N particles described without quantum statistics. Thus, Bose–Einstein statistics for a material particle is not a mechanism to bypass physical restrictions on the density of the corresponding substance, and superfluid liquid helium has the density comparable to the density of ordinary liquid matter. Spatially-delocalized states also permit for a low momentum according to uncertainty principle, hence for low kinetic energy; this is why superfluidity and superconductivity are usually observed in low temperatures.

Photons do not interact with themselves and hence do not experience this difference in states where to crowd.

GAUGE BOSON

In particle physics, a gauge boson is a force carrier, a bosonic particle that carries any of the fundamental interactions of nature, commonly called forces. Elementary particles, whose interactions are described by a gauge theory, interact with each other by the exchange of gauge bosons—usually as virtual particles.

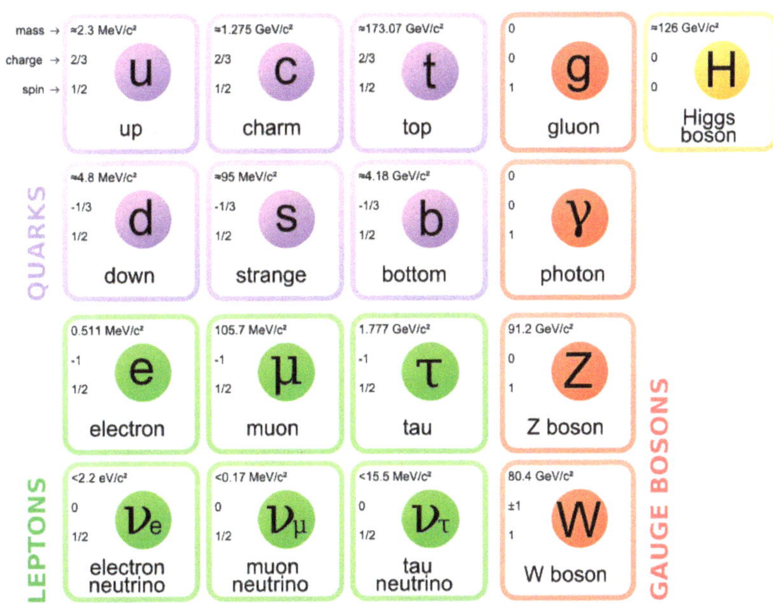

The Standard Model of elementary particles, with the gauge bosons in the fourth column in red.

All known gauge bosons have a spin of 1. Therefore, all known gauge bosons are vector bosons.

Gauge bosons are different from the other kinds of bosons: first, fundamental scalar bosons (the Higgs boson); second, mesons, which are composite bosons, made of quarks; third, larger composite, non-force-carrying bosons, such as certain atoms.

Gauge bosons in the Standard Model

The Standard Model of particle physics recognizes four kinds of gauge bosons: photons, which carry the electromagnetic interaction; W and Z bosons, which carry the weak interaction; and gluons, which carry the strong interaction.

Isolated gluons do not occur because they are color-charged and subject to color confinement.

Multiplicity of Gauge Bosons

In a quantized gauge theory, gauge bosons are quanta of the gauge fields. Consequently, there are as many gauge bosons as there are generators of the gauge field. In quantum electrodynamics, the gauge group is U(1); in this simple case, there is only one gauge boson. In quantum chromodynamics, the more complicated group SU(3) has eight generators, corresponding to the eight gluons. The three W and Z bosons correspond (roughly) to the three generators of SU(2) in GWS theory.

Massive Gauge Bosons

For technical reasons involving gauge invariance, gauge bosons are described mathematically by field equations for massless particles. Therefore, at a naïve theoretical level all gauge bosons are required to be massless, and the forces that they describe are required to be long-ranged. The conflict between this idea and experimental evidence that the weak and strong interactions have a very short range requires further theoretical insight.

According to the Standard Model, the W and Z bosons gain mass via the Higgs mechanism. In the Higgs mechanism, the four gauge bosons (of SU(2)×U(1) symmetry) of the unified electroweak interaction couple to a Higgs field. This field undergoes spontaneous symmetry breaking due to the shape of its interaction potential. As a result, the universe is permeated by a nonzero Higgs vacuum expectation value (VEV). This VEV couples to three of the electroweak gauge bosons (the Ws and Z), giving them mass; the remaining gauge boson remains massless (the photon). This theory also predicts the existence of a scalar Higgs boson, which has been observed in experiments at the LHC.

Beyond the Standard Model

Grand Unification Theories

The Georgi-Glashow model predicts additional gauge bosons named X and Y bosons.

The hypothetical X and Y bosons direct interactions between quarks and leptons, hence violating conservation of baryon number and causing proton decay. Such bosons would be even more massive than W and Z bosons due to symmetry breaking. Analysis of data collected from such sources as the Super-Kamiokande neutrino detector has yielded no evidence of X and Y bosons.

Gravitons

The fourth fundamental interaction, gravity, may also be carried by a boson, called the graviton. In the absence of experimental evidence and a mathematically coherent theory of quantum gravity, it is unknown whether this would be a gauge boson or not. The role of gauge invariance in general relativity is played by a similar symmetry: diffeomorphism invariance.

W' and Z' Bosons

W' and Z' bosons refer to hypothetical new gauge bosons (named in analogy with the Standard Model W and Z bosons).

Photon

A photon is an elementary particle, the quantum of all forms of electromagnetic radiation including light. It is the force carrier for electromagnetic force, even when static via virtual photons. The photon has zero rest mass and as a result, the interactions of this force with matter at long distance are observable at the microscopic and macroscopic levels. Like all elementary particles, photons are currently best explained by quantum mechanics and exhibit wave–particle duality, exhibiting properties of both waves and particles. For example, a single photon may be refracted by a lens and exhibit wave interference with itself, and it can behave as a particle with definite and finite measurable position and momentum. The photon's wave and quanta qualities are two observable aspects of a single phenomenon, and cannot be described by any mechanical model; a representation of this dual property of light, which assumes certain points on the wavefront to be the seat of the energy, is not possible. The quanta in a light wave cannot be spatially localized. Some defined physical parameters of a photon are listed.

The modern concept of the photon was developed gradually by Albert Einstein in the early 20th century to explain experimental observations that did not fit the classical wave model of light. The benefit of the photon model was that it accounted for the frequency dependence of light's energy, and explained the ability of matter and electromagnetic radiation to be in thermal equilibrium. The photon model accounted for anomalous observations, including the properties of black-body radiation, that others (notably Max Planck) had tried to explain using semiclassical models. In that model, light was described by Maxwell's equations, but material objects emitted and absorbed light in quantized amounts (i.e., they change energy only by certain particular discrete amounts). Although these semiclassical models contributed to the development of quantum mechanics, many further experiments beginning with the phenomenon of Compton scattering of single photons by electrons, validated Einstein's hypothesis that light itself is quantized. In

1926 the optical physicist Frithiof Wolfers and the chemist Gilbert N. Lewis coined the name photon for these particles. After Arthur H. Compton won the Nobel Prize in 1927 for his scattering studies, most scientists accepted that light quanta have an independent existence, and the term photon was accepted.

In the Standard Model of particle physics, photons and other elementary particles are described as a necessary consequence of physical laws having a certain symmetry at every point in spacetime. The intrinsic properties of particles, such as charge, mass and spin, are determined by this gauge symmetry. The photon concept has led to momentous advances in experimental and theoretical physics, including lasers, Bose–Einstein condensation, quantum field theory, and the probabilistic interpretation of quantum mechanics. It has been applied to photochemistry, high-resolution microscopy, and measurements of molecular distances. Recently, photons have been studied as elements of quantum computers, and for applications in optical imaging and optical communication such as quantum cryptography.

Nomenclature

In 1900, the German physicist Max Planck was studying black-body radiation and suggested that the energy carried by electromagnetic waves could only be released in "packets" of energy. In his 1901 article in Annalen der Physik he called these packets "energy elements". The word quanta (singular quantum, Latin for how much) was used before 1900 to mean particles or amounts of different quantities, including electricity. In 1905, Albert Einstein suggested that electromagnetic waves could only exist as discrete wave-packets. He called such a wave-packet the light quantum (German: das Lichtquant). The name photon derives from the Greek word for light, φῶς (transliterated phôs). Arthur Compton used photon in 1928, referring to Gilbert N. Lewis. The same name was used earlier, by the American physicist and psychologist Leonard T. Troland, who coined the word in 1916, in 1921 by the Irish physicist John Joly, in 1924 by the French physiologist René Wurmser (1890-1993) and in 1926 by the French physicist Frithiof Wolfers (1891-1971). The name was suggested initially as a unit related to the illumination of the eye and the resulting sensation of light and was used later in a physiological context. Although Wolfers's and Lewis's theories were contradicted by many experiments and never accepted, the new name was adopted very soon by most physicists after Compton used it.

In physics, a photon is usually denoted by the symbol γ (the Greek letter gamma). This symbol for the photon probably derives from gamma rays, which were discovered in 1900 by Paul Villard, named by Ernest Rutherford in 1903, and shown to be a form of electromagnetic radiation in 1914 by Rutherford and Edward Andrade. In chemistry and optical engineering, photons are usually symbolized by $h\nu$, the energy of a photon, where h is Planck's constant and the Greek letter ν (nu) is the photon›s frequency. Much less commonly, the photon can be symbolized by hf, where its frequency is denoted by f.

Physical Properties

A photon is massless, has no electric charge, and is a stable particle. A photon has two possible polarization states. In the momentum representation of the photon, which is preferred in quantum field theory, a photon is described by its wave vector, which determines its wavelength λ and its direction of propagation. A photon›s wave vector may not be zero and can be represented either as a spatial 3-vector or as a (relativistic) four-vector; in the latter case it belongs to the light cone (pictured). Different signs of the four-vector denote different circular polarizations, but in the 3-vector representation one should account for the polarization state separately; it actually is a spin quantum number. In both cases the space of possible wave vectors is three-dimensional.

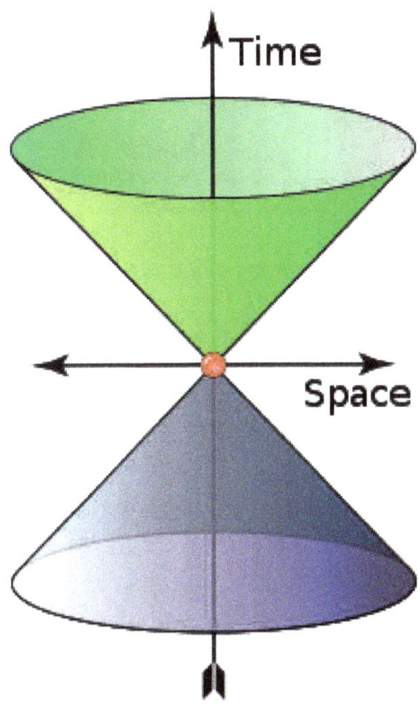

The cone shows possible values of wave 4-vector of a photon. The "time" axis gives the angular frequency (rad·s^{-1}) and the "space" axes represent the angular wavenumber (rad·m^{-1}). Green and indigo represent left and right polarization

The photon is the gauge boson for electromagnetism, and therefore all other quantum numbers of the photon (such as lepton number, baryon number, and flavor quantum numbers) are zero. Also, the photon does not obey the Pauli exclusion principle.

Photons are emitted in many natural processes. For example, when a charge is accelerated it emits synchrotron radiation. During a molecular, atomic or nuclear transition to a lower energy level, photons of various energy will be emitted, ranging from radio waves to gamma rays. A photon can also be emitted when a particle and its corresponding antiparticle are annihilated (for example, electron–positron annihilation).

In empty space, the photon moves at c (the speed of light) and its energy and momentum are related by E = pc, where p is the magnitude of the momentum vector **p**. This derives from the following relativistic relation, with m = 0:

$$E^2 = p^2c^2 + m^2c^4.$$

The energy and momentum of a photon depend only on its frequency (ν) or inversely, its wavelength (λ):

$$E = \hbar\omega = h\nu = \frac{hc}{\lambda}$$

$$\boldsymbol{p} = \hbar\boldsymbol{k},$$

where $\boldsymbol{k}$ is the wave vector (where the wave number $k = |\boldsymbol{k}| = 2\pi/\lambda$), $\omega = 2\pi\nu$ is the angular frequency, and $\hbar = h/2\pi$ is the reduced Planck constant.

Since $\boldsymbol{p}$ points in the direction of the photon's propagation, the magnitude of the momentum is

$$p = \hbar k = \frac{h\nu}{c} = \frac{h}{\lambda}.$$

The photon also carries a quantity called spin angular momentum that does not depend on its frequency. The magnitude of its spin is $\sqrt{2}\hbar$ and the component measured along its direction of motion, its helicity, must be $\pm\hbar$. These two possible helicities, called right-handed and left-handed, correspond to the two possible circular polarization states of the photon.

To illustrate the significance of these formulae, the annihilation of a particle with its antiparticle in free space must result in the creation of at least two photons for the following reason. In the center of momentum frame, the colliding antiparticles have no net momentum, whereas a single photon always has momentum (since, as we have seen, it is determined by the photon's frequency or wavelength, which cannot be zero). Hence, conservation of momentum (or equivalently, translational invariance) requires that at least two photons are created, with zero net momentum. (However, it is possible if the system interacts with another particle or field for the annihilation to produce one photon, as when a positron annihilates with a bound atomic electron, it is possible for only one photon to be emitted, as the nuclear Coulomb field breaks translational symmetry.)The energy of the two photons, or, equivalently, their frequency, may be determined from conservation of four-momentum. Seen another way, the photon can be considered as its own antiparticle. The reverse process, pair production, is the dominant mechanism by which high-energy photons such as gamma rays lose energy while passing through matter. That process is the reverse of "annihilation to one photon" allowed in the electric field of an atomic nucleus.

The classical formulae for the energy and momentum of electromagnetic radiation can be re-expressed in terms of photon events. For example, the pressure of electromagnetic radiation on an object derives from the transfer of photon momentum per unit time and unit area to that object, since pressure is force per unit area and force is the change in momentum per unit time.

Each photon carries two distinct and independent forms of angular momentum of light. The spin angular momentum of light of a particular photon is always either $+\hbar$ or $-\hbar$. The light orbital angular momentum of a particular photon can be any integer N, including zero.

Experimental Checks on Photon Mass

Current commonly accepted physical theories imply or assume the photon to be strictly massless. If the photon is not a strictly massless particle, it would not move at the exact speed of light, c in vacuum. Its speed would be lower and depend on its frequency. Relativity would be unaffected by this; the so-called speed of light, c, would then not be the actual speed at which light moves, but a constant of nature which is the upper bound on speed that any object could theoretically attain in space-time. Thus, it would still be the speed of space-time ripples (gravitational waves and gravitons), but it would not be the speed of photons.

If a photon did have non-zero mass, there would be other effects as well. Coulomb's law would be modified and the electromagnetic field would have an extra physical degree of freedom. These effects yield more sensitive experimental probes of the photon mass than the frequency dependence of the speed of light. If Coulomb's law is not exactly valid, then that would allow the presence of an electric field to exist within a hollow conductor when it is subjected to an external electric field. This thus allows one to test Coulomb's law to very high precision. A null result of such an experiment has set a limit of $m \lesssim 10^{-14}$ eV/c^2.

Sharper upper limits on the speed of light have been obtained in experiments designed to detect effects caused by the galactic vector potential. Although the galactic vector potential is very large because the galactic magnetic field exists on very great length scales, only the magnetic field would be observable if the photon is massless. In the case that the photon has mass, the mass term $\frac{1}{2}m^2 A_\mu A^\mu$ would affect the galactic plasma. The fact that no such effects are seen implies an upper bound on the photon mass of m < 3×10−27 eV/c2. The galactic vector potential can also be probed directly by measuring the torque exerted on a magnetized ring. Such methods were used to obtain the sharper upper limit of 10−18eV/c2 (the equivalent of 1.07×10−27 atomic mass units) given by the Particle Data Group.

These sharp limits from the non-observation of the effects caused by the galactic vector potential have been shown to be model dependent. If the photon mass is generated via the Higgs mechanism then the upper limit of m≲10−14 eV/c2 from the test of Coulomb's law is valid.

Photons inside superconductors do develop a nonzero effective rest mass; as a result, electromagnetic forces become short-range inside superconductors.

Historical Development

In most theories up to the eighteenth century, light was pictured as being made up of particles. Since particle models cannot easily account for the refraction, diffraction and birefringence of light, wave theories of light were proposed by René Descartes (1637), Robert Hooke (1665), and Christiaan Huygens (1678); however, particle models remained dominant, chiefly due to the influence of Isaac Newton.

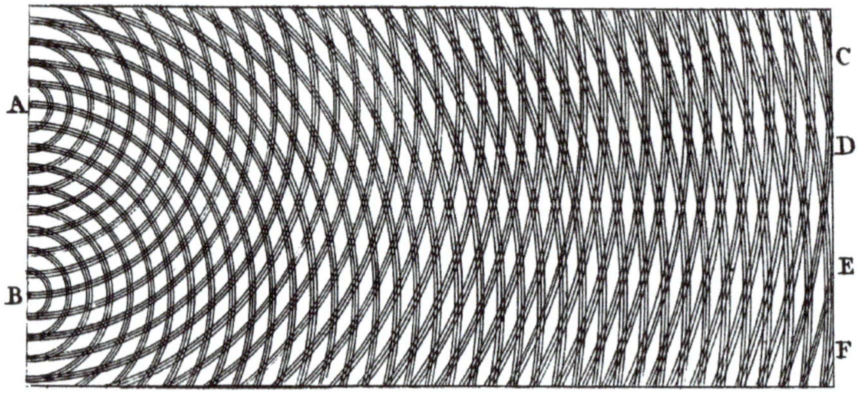

Thomas Young's double-slit experiment in 1801 showed that light can act as a wave, helping to invalidate early particle theories of light.

In the early nineteenth century, Thomas Young and August Fresnel clearly demonstrated the interference and diffraction of light and by 1850 wave models were generally accepted. In 1865, James Clerk Maxwell's prediction that light was an electromagnetic wave—which was confirmed experimentally in 1888 by Heinrich Hertz's detection of radio waves—seemed to be the final blow to particle models of light.

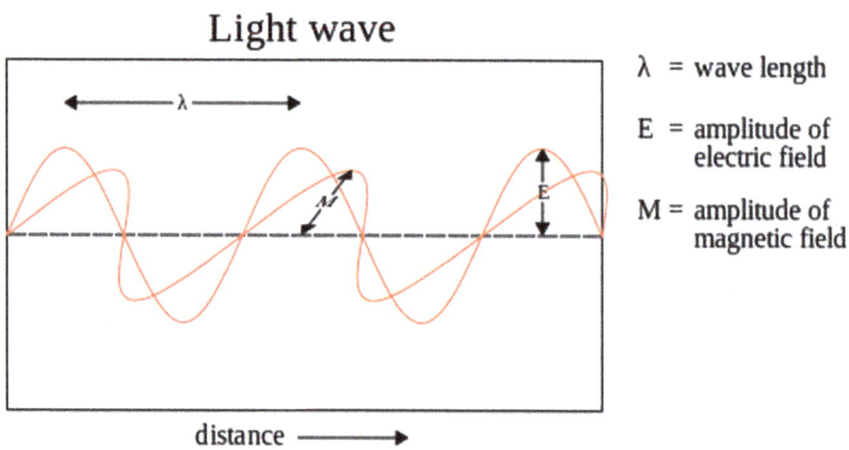

In 1900, Maxwell's theoretical model of light as oscillating electric and magnetic fields seemed complete. However, several observations could not be explained by any wave model of electromagnetic radiation, leading to the idea that light-energy was packaged into quanta described by E=hv. Later experiments showed that these light-quanta also carry momentum and, thus, can be considered particles: the photon concept was born, leading to a deeper understanding of the electric and magnetic fields themselves.

The Maxwell wave theory, however, does not account for all properties of light. The Maxwell theory predicts that the energy of a light wave depends only on its intensity, not on its frequency; nevertheless, several independent types of experiments show that the energy imparted by light to atoms depends only on the light's frequency, not on its intensity. For example, some chemical reactions are provoked only by light of frequency higher than a certain threshold; light of frequency lower than the threshold, no matter how intense, does not initiate the reaction. Similarly, electrons can be ejected from a metal plate by shining light of sufficiently high frequency on it (the photoelectric effect); the energy of the ejected electron is related only to the light's frequency, not to its intensity.

At the same time, investigations of blackbody radiation carried out over four decades (1860–1900) by various researchers culminated in Max Planck's hypothesis that the energy of any system that absorbs or emits electromagnetic radiation of frequency ν is an integer multiple of an energy quantum $E = h\nu$. As shown by Albert Einstein, some form of energy quantization must be assumed to account for the thermal equilibrium observed between matter and electromagnetic radiation; for this explanation of the photoelectric effect, Einstein received the 1921 Nobel Prize in physics.

Since the Maxwell theory of light allows for all possible energies of electromagnetic radiation, most physicists assumed initially that the energy quantization resulted from some unknown constraint on the matter that absorbs or emits the radiation. In 1905, Einstein was the first to propose that energy quantization was a property of electromagnetic radiation itself. Although he accepted the validity of Maxwell's theory, Einstein pointed out that many anomalous experiments could be explained if the energy of a Maxwellian light wave were localized into point-like quanta that move independently of one another, even if the wave itself is spread continuously over space. In 1909 and 1916, Einstein showed that, if Planck's law of black-body radiation is accepted, the energy quanta must also carry momentum $p = h/\lambda$, making them full-fledged particles. This photon momentum was observed experimentally by Arthur Compton, for which he received the Nobel Prize in 1927. The pivotal question was then: how to unify Maxwell›s wave theory of light with its experimentally observed particle nature? The answer to this question occupied Albert Einstein for the rest of his life, and was solved in quantum electrodynamics and its successor, the Standard Model.

Einstein's Light Quantum

Unlike Planck, Einstein entertained the possibility that there might be actual physical quanta of light—what we now call photons. He noticed that a light quantum with energy proportional to its frequency would explain a number of troubling puzzles and paradoxes, including an unpublished law by Stokes, the ultraviolet catastrophe, and the photoelectric effect. Stokes's law said simply that the frequency of fluorescent light cannot be greater than the frequency of the light (usually ultraviolet) inducing it. Einstein eliminated the ultraviolet catastrophe by imagining a gas of photons behaving like a gas of electrons that he had previously considered. He was advised by a colleague to be careful how he wrote up this paper, in order to not challenge Planck, a powerful figure in physics, too directly, and indeed the warning was justified, as Planck never forgave him for writing it.

Early Objections

Einstein's 1905 predictions were verified experimentally in several ways in the first two decades of the 20th century, as recounted in Robert Millikan's Nobel lecture. However, before Compton's experiment showed that photons carried momentum proportional to their wave number (1922), most physicists were reluctant to believe that electromagnetic radiation itself might be particulate. (See, for example, the Nobel lectures of Wien, Planck and Millikan.) Instead, there was a widespread belief that energy quantization resulted from some unknown constraint on the matter that absorbed or emitted radiation. Attitudes changed over time. In part, the change can be traced to experiments such as Compton scattering, where it was much more difficult not to ascribe quantization to light itself to explain the observed results.

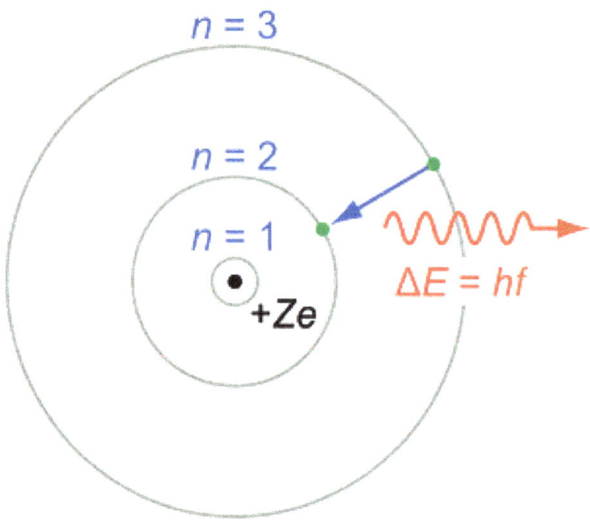

Up to 1923, most physicists were reluctant to accept that light itself was quantized. Instead, they tried to explain photon behavior by quantizing only matter, as in the Bohr model of the hydrogen atom (shown here). Even though these semiclassical models were only a first approximation, they were accurate for simple systems and they led to quantum mechanics.

Even after Compton's experiment, Niels Bohr, Hendrik Kramers and John Slater made one last attempt to preserve the Maxwellian continuous electromagnetic field model of light, the so-called BKS model. To account for the data then available, two drastic hypotheses had to be made:

- Energy and momentum are conserved only on the average in interactions between matter and radiation, but not in elementary processes such as absorption and emission. This allows one to reconcile the discontinuously changing energy of the atom (the jump between energy states) with the continuous release of energy as radiation.
- Causality is abandoned. For example, spontaneous emissions are merely emissions

stimulated by a "virtual" electromagnetic field.

However, refined Compton experiments showed that energy–momentum is conserved extraordinarily well in elementary processes; and also that the jolting of the electron and the generation of a new photon in Compton scattering obey causality to within 10 ps. Accordingly, Bohr and his co-workers gave their model "as honorable a funeral as possible". Nevertheless, the failures of the BKS model inspired Werner Heisenberg in his development of matrix mechanics.

A few physicists persisted in developing semiclassical models in which electromagnetic radiation is not quantized, but matter appears to obey the laws of quantum mechanics. Although the evidence from chemical and physical experiments for the existence of photons was overwhelming by the 1970s, this evidence could not be considered as absolutely definitive; since it relied on the interaction of light with matter, and a sufficiently complete theory of matter could in principle account for the evidence. Nevertheless, all semiclassical theories were refuted definitively in the 1970s and 1980s by photon-correlation experiments. Hence, Einstein's hypothesis that quantization is a property of light itself is considered to be proven.

Wave–Particle Duality and Uncertainty Principles

Photons, like all quantum objects, exhibit wave-like and particle-like properties. Their dual wave–particle nature can be difficult to visualize. The photon displays clearly wave-like phenomena such as diffraction and interference on the length scale of its wavelength. For example, a single photon passing through a double-slit experiment exhibits interference phenomena but only if no measure was made at the slit. A single photon passing through a double-slit experiment lands on the screen with a probability distribution given by its interference pattern determined by Maxwell's equations.

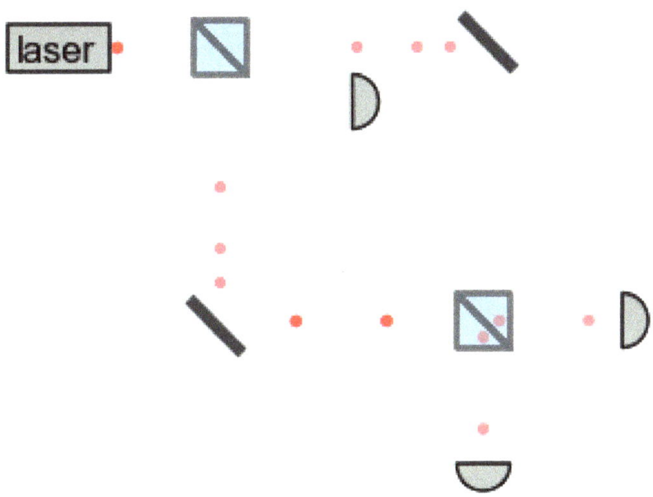

Photons in a Mach–Zehnder interferometer exhibit wave-like interference and particle-like detection at single-photon detectors.

However, experiments confirm that the photon is not a short pulse of electromagnetic radiation; it does not spread out as it propagates, nor does it divide when it encounters a beam splitter. Rather, the photon seems to be a point-like particle since it is absorbed or emitted as a whole by arbitrarily small systems, systems much smaller than its wavelength, such as an atomic nucleus ($\approx 10^{-15}$ m across) or even the point-like electron. Nevertheless, the photon is not a point-like particle whose trajectory is shaped probabilistically by the electromagnetic field, as conceived by Einstein and others; that hypothesis was also refuted by the photon-correlation experiments cited above. According to our present understanding, the electromagnetic field itself is produced by photons, which in turn result from a local gauge symmetry and the laws of quantum field theory.

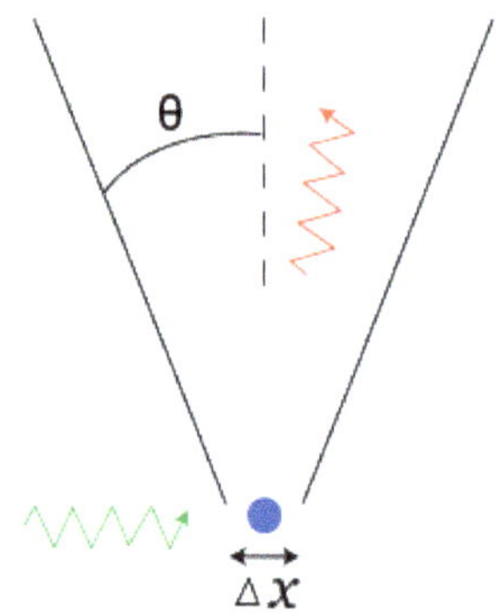

Heisenberg's thought experiment for locating an electron (shown in blue) with a high-resolution gamma-ray microscope. The incoming gamma ray (shown in green) is scattered by the electron up into the microscope's aperture angle θ. The scattered gamma ray is shown in red. Classical optics shows that the electron position can be resolved only up to an uncertainty Δx that depends on θ and the wavelength λ of the incoming light.

A key element of quantum mechanics is Heisenberg›s uncertainty principle, which forbids the simultaneous measurement of the position and momentum of a particle along the same direction. Remarkably, the uncertainty principle for charged, material particles requires the quantization of light into photons, and even the frequency dependence of the photon's energy and momentum.

An elegant illustration of the uncertainty principle is Heisenberg's thought experiment for locating an electron with an ideal microscope. The position of the electron can be determined to within the resolving power of the microscope, which is given by a formula from classical optics

$$\Delta x \sim \frac{\lambda}{\sin\theta}$$

where θ is the aperture angle of the microscope and λ is the wavelength of the light used to observe the electron. Thus, the position uncertainty Δx can be made arbitrarily small by reducing the wavelength λ. Even if the momentum of the electron is initially known, the light impinging on the electron will give it a momentum «kick» Δp of some unknown amount, rendering the momentum of the electron uncertain. If light were not quantized into photons, the uncertainty Δp could be made arbitrarily small by reducing the light's intensity. In that case, since the wavelength and intensity of light can be varied independently, one could simultaneously determine the position and momentum to arbitrarily high accuracy, violating the uncertainty principle. By contrast, Einstein's formula for photon momentum preserves the uncertainty principle; since the photon is scattered anywhere within the aperture, the uncertainty of momentum transferred equals

$$\Delta p \sim p_{\text{photon}} \sin\theta = \frac{h}{\lambda} \sin\theta$$

giving the product $\Delta x \Delta p \sim h$, which is Heisenberg's uncertainty principle. Thus, the entire world is quantized; both matter and fields must obey a consistent set of quantum laws, if either one is to be quantized.

The analogous uncertainty principle for photons forbids the simultaneous measurement of the number n of photons (see Fock state and the Second quantization section below) in an electromagnetic wave and the phase ϕ of that wave

$$\Delta n \Delta \phi > 1$$

Both photons and electrons create analogous interference patterns when passed through a double-slit experiment. For photons, this corresponds to the interference of a Maxwell light wave whereas, for material particles (electron), this corresponds to the interference of the Schrödinger wave equation. Although this similarity might suggest that Maxwell's equations describing the photon's electromagnetic wave are simply Schrödinger's equation for photons, most physicists do not agree. For one thing, they are mathematically different; most obviously, Schrödinger's one equation for the electron solves for a complex field, whereas Maxwell's four equations solve for real fields. More generally, the normal concept of a Schrödinger probability wave function cannot be applied to photons. As photons are massless, they cannot be localized without being destroyed; technically, photons cannot have a position eigenstate $|\mathbf{r}\rangle$, and, thus, the normal Heisenberg uncertainty principle $\Delta x \Delta p > h/2$ does not pertain to photons. A few substitute wave functions have been suggested for the photon, but they have not come into general use. Instead, physicists generally accept the second-quantized theory of photons described below, quantum electrodynamics, in which photons are quantized excitations of electromagnetic modes.

Another interpretation, that avoids duality, is the De Broglie–Bohm theory, known also as the pilot-wave model. In that theory, the photon is both, wave and particle. "This idea seems to me so natural and simple, to resolve the wave-particle dilemma in such a clear and ordinary way, that it is a great mystery to me that it was so generally ignored", J.S.Bell.

Bose–Einstein Model of a Photon Gas

In 1924, Satyendra Nath Bose derived Planck's law of black-body radiation without using any electromagnetism, but rather by using a modification of coarse-grained counting of phase space. Einstein showed that this modification is equivalent to assuming that photons are rigorously identical and that it implied a "mysterious non-local interaction", now understood as the requirement for a symmetric quantum mechanical state. This work led to the concept of coherent states and the development of the laser. In the same papers, Einstein extended Bose's formalism to material particles (bosons) and predicted that they would condense into their lowest quantum state at low enough temperatures; this Bose–Einstein condensation was observed experimentally in 1995. It was later used by Lene Hau to slow, and then completely stop, light in 1999 and 2001.

The modern view on this is that photons are, by virtue of their integer spin, bosons (as opposed to fermions with half-integer spin). By the spin-statistics theorem, all bosons obey Bose–Einstein statistics (whereas all fermions obey Fermi–Dirac statistics).

Stimulated and Spontaneous Emission

In 1916, Albert Einstein showed that Planck's radiation law could be derived from a semi-classical, statistical treatment of photons and atoms, which implies a link between the rates at which atoms emit and absorb photons. The condition follows from the assumption that functions of the emission and absorption of radiation by the atoms are independent of each other, and that thermal equilibrium is made by way of the radiation's interaction with the atoms.

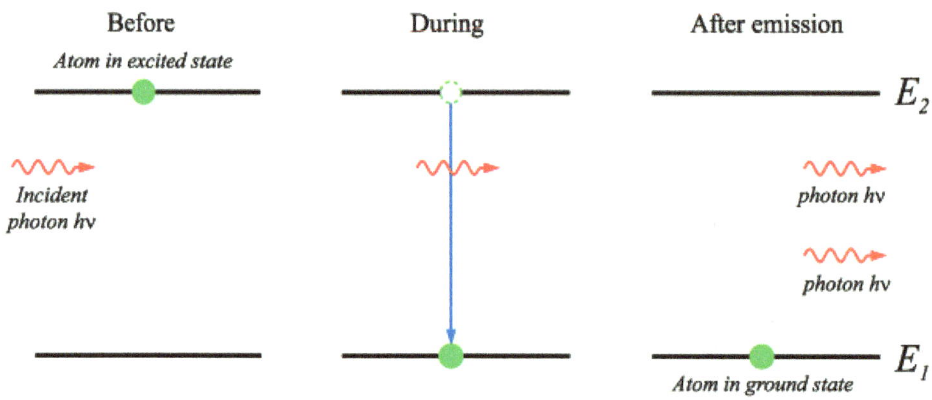

Stimulated emission (in which photons "clone" themselves) was predicted by Einstein in his kinetic analysis, and led to the development of the laser. Einstein's derivation inspired further developments in the quantum treatment of light, which led to the statistical interpretation of quantum mechanics.

Consider a cavity in thermal equilibrium with all parts of itself and filled with electromagnetic radiation and that the atoms can emit and absorb that radiation. Thermal equilibrium requires that the energy density $\rho(\nu)$ of photons with frequency ν (which is proportional to their number density) is, on average, constant in time; hence, the rate at which photons of any particular frequency are emitted must equal the rate at which they absorb them.

Einstein began by postulating simple proportionality relations for the different reaction rates involved. In his model, the rate R_{ji} for a system to absorb a photon of frequency V and transition from a lower energy E_j to a higher energy E_i is proportional to the number N_j of atoms with energy E_j and to the energy density P(v) of ambient photons of that frequency,

$$R_{ji} = N_j B_{ji} \rho(\nu)$$

where B_{ji} is the rate constant for absorption. For the reverse process, there are two possibilities: spontaneous emission of a photon, or the emission of a photon initiated by the interaction of the atom with a passing photon and the return of the atom to the lower-energy state. Following Einstein's approach, the corresponding rate R_{ij} for the emission of photons of frequency ν and transition from a higher energy E_i to a lower energy E_j is

$$R_{ij} = N_i A_{ij} + N_i B_{ij} \rho(\nu)$$

where A_{ij} is the rate constant for emitting a photon spontaneously, and B_{ij} is the rate constant for emissions in response to ambient photons (induced or stimulated emission). In thermodynamic equilibrium, the number of atoms in state i and those in state j must, on average, be constant; hence, the rates R_{ji} and R_{ij} must be equal. Also, by arguments analogous to the derivation of Boltzmann statistics, the ratio of N_i and N_j is $g_i/g_j \exp{(E_j - E_i)/(kT)}$, where $g_{i,j}$ are the degeneracy of the state i and that of j, respectively, E_{ij} their energies, k the Boltzmann constant and T the system's temperature. From this, it is readily derived that $g_i B_{ij} = g_j B_{ji}$ and

$$A_{ij} = \frac{8\pi h \nu^3}{c^3} B_{ij}.$$

The A and Bs are collectively known as the Einstein coefficients.

Einstein could not fully justify his rate equations, but claimed that it should be possible to calculate the coefficients A_{ij}, B_{ji}, B_{ij} once physicists had obtained "mechanics and electrodynamics modified to accommodate the quantum hypothesis". In fact, in 1926, Paul Dirac derived the B_{ij} rate constants by using a semiclassical approach, and, in 1927, succeeded in deriving all the rate constants from first principles within the framework of quantum theory. Dirac's work was the foundation of quantum electrodynamics, i.e., the quantization of the electromagnetic field itself. Dirac's approach is also called second quantization or quantum field theory; earlier quantum mechanical treatments only treat material particles as quantum mechanical, not the electromagnetic field.

Einstein was troubled by the fact that his theory seemed incomplete, since it did not determine the direction of a spontaneously emitted photon. A probabilistic nature of light-particle motion was first considered by Newton in his treatment of birefringence and, more generally, of the splitting of light beams at interfaces into a transmitted beam and a reflected beam. Newton hypothesized that hidden variables in the light particle determined which of the two paths a single photon would take. Similarly, Einstein hoped for a more complete theory that would leave nothing to chance, beginning his separation from quantum mechanics. Ironically, Max Born's probabilistic interpretation of the wave function was inspired by Einstein's later work searching for a more complete theory.

Second Quantization and High Energy Photon Interactions

In 1910, Peter Debye derived Planck's law of black-body radiation from a relatively simple assumption. He correctly decomposed the electromagnetic field in a cavity into its Fourier modes, and assumed that the energy in any mode was an integer multiple of $h\nu$, where ν is the frequency of the electromagnetic mode. Planck's law of black-body radiation follows immediately as a geometric sum. However, Debye's approach failed to give the correct formula for the energy fluctuations of blackbody radiation, which were derived by Einstein in 1909.

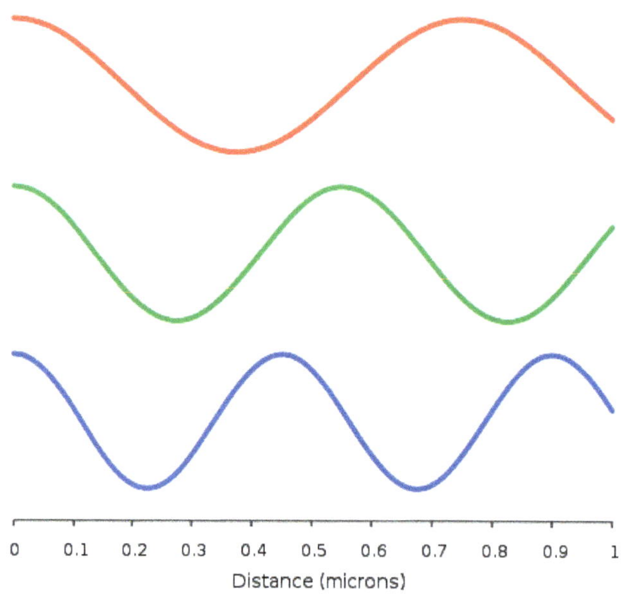

Different electromagnetic modes (such as those depicted here) can be treated as independent simple harmonic oscillators. A photon corresponds to a unit of energy E=hv in its electromagnetic mode.

In 1925, Born, Heisenberg and Jordan reinterpreted Debye's concept in a key way. As may be shown classically, the Fourier modes of the electromagnetic field—a complete set of electromagnetic plane waves indexed by their wave vector k and polarization state—are equivalent to a set of uncoupled simple harmonic oscillators. Treated quantum mechanically, the energy levels of such oscillators are known to be $E = nh\nu$, where ν is the oscillator frequency. The key new step was to identify an electromagnetic mode with energy $E = nh\nu$ as a state with n photons, each of energy $h\nu$. This approach gives the correct energy fluctuation formula.

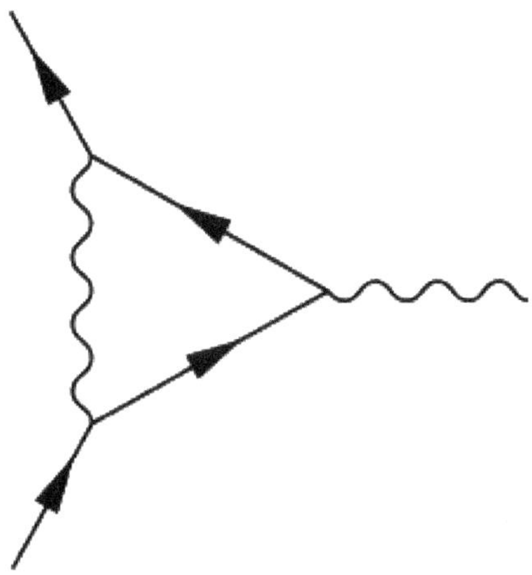

In quantum field theory, the probability of an event is computed by summing the probability amplitude (a complex number) for all possible ways in which the event can occur, as in the Feynman diagram shown here; the probability equals the square of the modulus of the total amplitude.

Dirac took this one step further. He treated the interaction between a charge and an electromagnetic field as a small perturbation that induces transitions in the photon states, changing the numbers of photons in the modes, while conserving energy and momentum overall. Dirac was able to derive Einstein's A_{ij} and B_{ij} coefficients from first principles, and showed that the Bose–Einstein statistics of photons is a natural consequence of quantizing the electromagnetic field correctly (Bose's reasoning went in the opposite direction; he derived Planck's law of black-body radiation by assuming B–E statistics). In Dirac's time, it was not yet known that all bosons, including photons, must obey Bose–Einstein statistics.

Dirac's second-order perturbation theory can involve virtual photons, transient intermediate states of the electromagnetic field; the static electric and magnetic interactions are mediated by such virtual photons. In such quantum field theories, the probability amplitude of observable events is calculated by summing over all possible intermediate steps, even ones that are unphysical; hence, virtual photons are not constrained to satisfy $E = pc$, and may have extra polarization states; depending on the gauge used, virtual photons may have three or four polarization states, instead of the two states of real photons. Although these transient virtual photons can never be observed, they contribute measurably to the probabilities of observable events. Indeed, such second-order and higher-order perturbation calculations can give apparently infinite contributions to the sum. Such unphysical results are corrected for using the technique of renormalization.

Other virtual particles may contribute to the summation as well; for example, two photons may interact indirectly through virtual electron–positron pairs. In fact, such photon-photon scattering, as well as electron-photon scattering, is meant to be one of the modes of operations of the planned particle accelerator, the International Linear Collider.

In modern physics notation, the quantum state of the electromagnetic field is written as a Fock state, a tensor product of the states for each electromagnetic mode

$$|n_{k_0}\rangle \otimes |n_{k_1}\rangle \otimes \cdots \otimes |n_{k_n}\rangle \ldots$$

where $|n_{k_i}\rangle$ represents the state in which n_{k_i} photons are in the mode k_i. In this notation, the creation of a new photon in mode k_i (e.g., emitted from an atomic transition) is written as $|n_{k_i}\rangle \rightarrow |n_{k_i}+1\rangle$. This notation merely expresses the concept of Born, Heisenberg and Jordan described above, and does not add any physics.

The Hadronic Properties of the Photon

Measurements of the interaction between energetic photons and hadrons show that the interaction is much more intense than expected by the interaction of merely photons with the hadron's electric charge. Furthermore, the interaction of energetic photons with protons is similar to the interaction of photons with neutrons in spite of the fact that the electric charge structures of protons and neutrons are substantially different. A theory called Vector Meson Dominance (VMD) was developed to explain this effect. According to VMD, the photon is a superposition of the pure electromagnetic photon which interacts only with electric charges and vector meson.However, if experimentally probed at very short distances, the intrinsic structure of the photon is recognized as a flux of quark and gluon components, quasi-free according to asymptotic freedom in QCD and described by the photon structure function. A comprehensive comparison of data with theoretical predictions is presented in a recent review.

The Photon as a Gauge Boson

The electromagnetic field can be understood as a gauge field, i.e., as a field that results from requiring that a gauge symmetry holds independently at every position in spacetime. For the electromagnetic field, this gauge symmetry is the Abelian U(1) symmetry of complex numbers of absolute value 1, which reflects the ability to vary the phase of a complex field without affecting observables or real valued functions made from it, such as the energy or the Lagrangian.

The quanta of an Abelian gauge field must be massless, uncharged bosons, as long as the symmetry is not broken; hence, the photon is predicted to be massless, and to have zero electric charge and integer spin. The particular form of the electromagnetic interaction specifies that the photon must have spin ±1; thus, its helicity must be $\pm\hbar$. These two spin components correspond to the classical concepts of right-handed and left-handed circularly polarized light. However, the transient virtual photons of quantum electrodynamics may also adopt unphysical polarization states.

In the prevailing Standard Model of physics, the photon is one of four gauge bosons in the electroweak interaction; the other three are denoted W^+, W^- and Z^0 and are responsible for the weak interaction. Unlike the photon, these gauge bosons have mass, owing to a mechanism that breaks their SU(2) gauge symmetry. The unification of the photon with W and Z gauge bosons in the electroweak interaction was accomplished by Sheldon Glashow, Abdus Salam and Steven Weinberg, for which they were awarded the 1979 Nobel Prize in physics. Physicists continue to hypothesize grand unified theories that connect these four gauge bosons with the eight gluon gauge bosons of quantum chromodynamics; however, key predictions of these theories, such as proton decay, have not been observed experimentally.

Contributions to the Mass of a System

The energy of a system that emits a photon is decreased by the energy E of the photon as measured in the rest frame of the emitting system, which may result in a reduction in mass in the amount E/c^2. Similarly, the mass of a system that absorbs a photon is increased by a corresponding amount. As an application, the energy balance of nuclear reactions involving photons is commonly written in terms of the masses of the nuclei involved, and terms of the form E/c^2 for the gamma photons (and for other relevant energies, such as the recoil energy of nuclei).

This concept is applied in key predictions of quantum electrodynamics (QED). In that theory, the mass of electrons (or, more generally, leptons) is modified by including the mass contributions of virtual photons, in a technique known as renormalization. Such "radiative corrections" contribute to a number of predictions of QED, such as the magnetic dipole moment of leptons, the Lamb shift, and the hyperfine structure of bound lepton pairs, such as muonium and positronium.

Since photons contribute to the stress–energy tensor, they exert a gravitational attraction on other objects, according to the theory of general relativity. Conversely, photons are themselves affected by gravity; their normally straight trajectories may be bent by warped spacetime, as in gravitational lensing, and their frequencies may be lowered by moving to a higher gravitational potential, as in the Pound–Rebka experiment. However, these effects are not specific to photons; exactly the same effects would be predicted for classical electromagnetic waves.

Photons in Matter

Light that travels through transparent matter does so at a lower speed than c, the speed of light in a vacuum. For example, photons engage in so many collisions on the way from the core of the sun that radiant energy can take about a million years to reach the surface; however, once in open space, a photon takes only 8.3 minutes to reach Earth. The factor by which the speed is decreased is called the refractive index of the material. In a classical wave picture, the slowing can be explained by the light inducing electric polarization in the matter, the polarized matter radiating new light, and that new light interfering with the original light wave to form a delayed wave. In a particle picture, the slowing can instead be described as a blending of the photon with quantum excitations of the matter to produce quasi-particles known as polariton (other quasi-particles are phonons and excitons); this polariton has a nonzero effective mass, which means that it cannot travel at c. Light of different frequencies may travel through matter at different speeds; this is called dispersion (not to be confused with scattering). In some cases, it can result in extremely slow speeds of light in matter. The effects of photon interactions with other quasi-particles may be observed directly in Raman scattering and Brillouin scattering.

Photons can also be absorbed by nuclei, atoms or molecules, provoking transitions between their energy levels. A classic example is the molecular transition of retinal

($C_{20}H_{28}O$), which is responsible for vision, as discovered in 1958 by Nobel laureate biochemist George Wald and co-workers. The absorption provokes a cis-trans isomerization that, in combination with other such transitions, is transduced into nerve impulses. The absorption of photons can even break chemical bonds, as in the photodissociation of chlorine; this is the subject of photochemistry.

Technological Applications

Photons have many applications in technology. These examples are chosen to illustrate applications of photons per se, rather than general optical devices such as lenses, etc. that could operate under a classical theory of light. The laser is an extremely important application and is discussed above under stimulated emission.

Individual photons can be detected by several methods. The classic photomultiplier tube exploits the photoelectric effect: a photon of sufficient energy strikes a metal plate and knocks free an electron, initiating an ever-amplifying avalanche of electrons. Semiconductor charge-coupled device chips use a similar effect: an incident photon generates a charge on a microscopic capacitor that can be detected. Other detectors such as Geiger counters use the ability of photons to ionize gas molecules contained in the device, causing a detectable change of conductivity of the gas.

Planck's energy formula $E = h\nu$ is often used by engineers and chemists in design, both to compute the change in energy resulting from a photon absorption and to determine the frequency of the light emitted from a given photon emission. For example, the emission spectrum of a gas-discharge lamp can be altered by filling it with (mixtures of) gases with different electronic energy level configurations.

Under some conditions, an energy transition can be excited by "two" photons that individually would be insufficient. This allows for higher resolution microscopy, because the sample absorbs energy only in the spectrum where two beams of different colors overlap significantly, which can be made much smaller than the excitation volume of a single beam. Moreover, these photons cause less damage to the sample, since they are of lower energy.

In some cases, two energy transitions can be coupled so that, as one system absorbs a photon, another nearby system "steals" its energy and re-emits a photon of a different frequency. This is the basis of fluorescence resonance energy transfer, a technique that is used in molecular biology to study the interaction of suitable proteins.

Several different kinds of hardware random number generators involve the detection of single photons. In one example, for each bit in the random sequence that is to be produced, a photon is sent to a beam-splitter. In such a situation, there are two possible outcomes of equal probability. The actual outcome is used to determine whether the next bit in the sequence is "0" or "1".

Recent Research

Much research has been devoted to applications of photons in the field of quantum optics. Photons seem well-suited to be elements of an extremely fast quantum computer, and the quantum entanglement of photons is a focus of research. Nonlinear optical processes are another active research area, with topics such as two-photon absorption, self-phase modulation, modulational instability and optical parametric oscillators. However, such processes generally do not require the assumption of photons per se; they may often be modeled by treating atoms as nonlinear oscillators. The nonlinear process of spontaneous parametric down conversion is often used to produce single-photon states. Finally, photons are essential in some aspects of optical communication, especially for quantum cryptography.

Gluon

Gluons are elementary particles that act as the exchange particles (or gauge bosons) for the strong force between quarks, analogous to the exchange of photons in the electromagnetic force between two charged particles. In lay terms, they "glue" quarks together, forming protons and neutrons.

In technical terms, gluons are vector gauge bosons that mediate strong interactions of quarks in quantum chromodynamics (QCD). Gluons themselves carry the color charge of the strong interaction. This is unlike the photon, which mediates the electromagnetic interaction but lacks an electric charge. Gluons therefore participate in the strong interaction in addition to mediating it, making QCD significantly harder to analyze than QED (quantum electrodynamics).

Properties

The gluon is a vector boson; like the photon, it has a spin of 1. While massive spin-1 particles have three polarization states, massless gauge bosons like the gluon have only two polarization states because gauge invariance requires the polarization to be transverse. In quantum field theory, unbroken gauge invariance requires that gauge bosons have zero mass (experiment limits the gluon's rest mass to less than a few meV/c^2). The gluon has negative intrinsic parity.

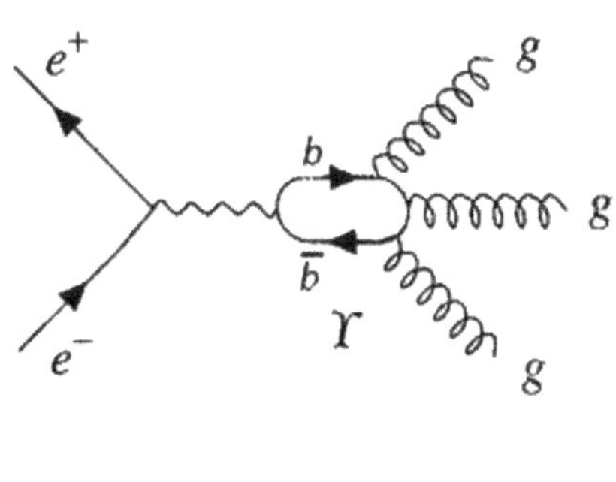

Diagram 2: $e^+e^- \rightarrow \Upsilon(9.46) \rightarrow 3g$

Numerology of Gluons

Unlike the single photon of QED or the three W and Z bosons of the weak interaction, there are eight independent types of gluon in QCD.

This may be difficult to understand intuitively. Quarks carry three types of color charge; antiquarks carry three types of anticolor. Gluons may be thought of as carrying both color and anticolor, but to correctly understand how they are combined, it is necessary to consider the mathematics of color charge in more detail.

Color Charge and Superposition

In quantum mechanics, the states of particles may be added according to the principle of superposition; that is, they may be in a "combined state" with a probability, if some particular quantity is measured, of giving several different outcomes. A relevant illustration in the case at hand would be a gluon with a color state described by:

$$(r\bar{b} + b\bar{r})/\sqrt{2}.$$

This is read as "red–antiblue plus blue–antired". (The factor of the square root of two is required for normalization, a detail that is not crucial to understand in this discussion.) If one were somehow able to make a direct measurement of the color of a gluon in this state, there would be a 50% chance of it having red-antiblue color charge and a 50% chance of blue-antired color charge.

Color Singlet States

It is often said that the stable strongly interacting particles (such as the proton and the neutron, i.e. hadrons) observed in nature are "colorless", but more precisely they are in a "color singlet" state, which is mathematically analogous to a spin singlet state. Such states allow interaction with other color singlets, but not with other color states; because long-range gluon interactions do not exist, this illustrates that gluons in the singlet state do not exist either.

The color singlet state is:

$$(r\bar{r} + b\bar{b} + g\bar{g})/\sqrt{3}.$$

In words, if one could measure the color of the state, there would be equal probabilities of it being red-antired, blue-antiblue, or green-antigreen.

Eight Gluon Colors

There are eight remaining independent color states, which correspond to the "eight types" or "eight colors" of gluons. Because states can be mixed together as discussed above, there are many ways of presenting these states, which are known as the "color octet". One commonly used list is:

$$
\begin{array}{ll}
(r\bar{b}+b\bar{r})/\sqrt{2} & -i(r\bar{b}-b\bar{r})/\sqrt{2} \\
(r\bar{g}+g\bar{r})/\sqrt{2} & -i(r\bar{g}-g\bar{r})/\sqrt{2} \\
(b\bar{g}+g\bar{b})/\sqrt{2} & -i(b\bar{g}-g\bar{b})/\sqrt{2} \\
(r\bar{r}-b\bar{b})/\sqrt{2} & (r\bar{r}+b\bar{b}-2g\bar{g})/\sqrt{6}.
\end{array}
$$

These are equivalent to the Gell-Mann matrices. The critical feature of these particular eight states is that they are linearly independent, and also independent of the singlet state; there is no way to add any combination of states to produce any other. (It is also impossible to add them to make rr, gg, or bb otherwise the forbidden singlet state could also be made.) There are many other possible choices, but all are mathematically equivalent, at least equally complex, and give the same physical results.

Group Theory Details

Technically, QCD is a gauge theory with SU(3) gauge symmetry. Quarks are introduced as spinors in N_f flavors, each in the fundamental representation (triplet, denoted **3**) of the color gauge group, SU(3). The gluons are vectors in the adjoint representation (octets, denoted **8**) of color SU(3). For a general gauge group, the number of force-carriers (like photons or gluons) is always equal to the dimension of the adjoint representation. For the simple case of SU(N), the dimension of this representation is $N^2 - 1$.

In terms of group theory, the assertion that there are no color singlet gluons is simply the statement that quantum chromodynamics has an SU(3) rather than a U(3) symmetry. There is no known a priori reason for one group to be preferred over the other, but as discussed above, the experimental evidence supports SU(3). The U(1) group for electromagnetic field combines with a slightly more complicated group known as SU(2) – S stands for "special" – which means the corresponding matrices have determinant 1 in addition to being unitary.

Confinement

Since gluons themselves carry color charge, they participate in strong interactions. These gluon-gluon interactions constrain color fields to string-like objects called "flux tubes", which exert constant force when stretched. Due to this force, quarks are confined within composite particles called hadrons. This effectively limits the range of the strong interaction to 1×10^{-15}meters, roughly the size of an atomic nucleus. Beyond a certain distance, the energy of the flux tube binding two quarks increases linearly. At a large enough distance, it becomes energetically more favorable to pull a quark-antiquark pair out of the vacuum rather than increase the length of the flux tube.

Gluons also share this property of being confined within hadrons. One consequence is that gluons are not directly involved in the nuclear forces between hadrons. The force mediators for these are other hadrons called mesons.

Although in the normal phase of QCD single gluons may not travel freely, it is predicted that there exist hadrons that are formed entirely of gluons — called glueballs.

There are also conjectures about other exotic hadrons in which real gluons (as opposed to virtual ones found in ordinary hadrons) would be primary constituents. Beyond the normal phase of QCD (at extreme temperatures and pressures), quark–gluon plasma forms. In such a plasma there are no hadrons; quarks and gluons become free particles.

Experimental Observations

Quarks and gluons (colored) manifest themselves by fragmenting into more quarks and gluons, which in turn hadronize into normal (colorless) particles, correlated in jets. As shown in 1978 summer conferences the PLUTO detector at the electron-positron collider DORIS (DESY) produced the first evidence that the hadronic decays of the very narrow resonance Y(9.46) could be interpreted as three-jet event topologies produced by three gluons. Later published analyses by the same experiment confirmed this interpretation and also the spin 1 nature of the gluon.

In summer 1979 at higher energies at the electron-positron collider PETRA (DESY) again three-jet topologies were observed, now interpreted as qq gluon bremsstrahlung, now clearly visible, by TASSO, MARK-J and PLUTO experiments (later in 1980 also by JADE). The spin 1 of the gluon was confirmed in 1980 by TASSO and PLUTO experiments. In 1991 a subsequent experiment at the LEP storage ring at CERN again confirmed this result.

The gluons play an important role in the elementary strong interactions between quarks and gluons, described by QCD and studied particularly at the electron-proton collider HERA at DESY. The number and momentum distribution of the gluons in the proton (gluon density) have been measured by two experiments, H1 and ZEUS, in the years 1996 till today (2012). The gluon contribution to the proton spin has been studied by the HERMES experiment at HERA. The gluon density in the proton (when behaving hadronically) also has been measured.

Color confinement is verified by the failure of free quark searches (searches of fractional charges). Quarks are normally produced in pairs (quark + antiquark) to compensate the quantum color and flavor numbers; however at Fermilab single production of top quarks has been shown (technically this still involves a pair production, but quark and antiquark are of different flavor). No glueball has been demonstrated.

Deconfinement was claimed in 2000 at CERN SPS in heavy-ion collisions, and it implies a new state of matter: quark–gluon plasma, less interacting than in the nucleus, almost as in a liquid. It was found at the Relativistic Heavy Ion Collider (RHIC) at Brookhaven in the years 2004–2010 by four contemporaneous experiments. A quark–gluon plasma state has been confirmed at the CERN Large Hadron Collider (LHC) by the three experiments ALICE, ATLAS and CMS in 2010.

The Continuous Electron Beam Accelerator Facility at Jefferson Lab, also called the Thomas Jefferson National Accelerator Facility, in Newport News, Virginia is one of 10 Department of Energy facilities doing research on gluons. The Virginia lab is competing with another facility in Long Island, New York, Brookhaven National Laboratory, for funds to build a new electron-ion collider.

W and Z Bosons

The W and Z bosons are together known as the weak or more generally as the intermediate vector bosons. These elementary particlesmediate the weak interaction; the respective symbols are W^+, W^-, and Z. The W boson has either a positive or negative electric charge of 1 elementary charge and are each other's antiparticles. The Z boson is electrically neutral and is its own antiparticle. The three particles have a spin of 1. The W bosons have a magnetic moment, but the Z has none. All three of these particles are very short-lived, with a half-life of about 3×10^{-25} s. Their experimental discovery was a triumph for what is now known as the Standard Model of particle physics.

Basic Properties

These bosons are among the heavyweights of the elementary particles. With masses of 80.4 GeV/c^2 and 91.2 GeV/c^2, respectively, the W and Z bosons are almost 100 times as large as the proton – heavier, even, than entire iron atoms. The masses of these bosons are significant because they act as the force carriers of a quite short-range fundamental force: their high masses thus limit the range of the weak nuclear force. By way of contrast, the electromagnetic force has an infinite range, because its force carrier, the photon, has zero mass, and the same is supposed of the hypothetical graviton.

All three bosons have particle spin s = 1. The emission of a W+ or W− boson either raises or lowers the electric charge of the emitting particle by one unit, and also alters the spin by one unit. At the same time, the emission or absorption of a W boson can change the type of the particle – for example changing a strange quark into an up quark. The neutral Z boson cannot change the electric charge of any particle, nor can it change any other of the so-called "charges" (such as strangeness, baryon number, charm, etc.). The emission or absorption of a Z boson can only change the spin, momentum, and energy of the other particle.

Weak Nuclear Force

The W and Z bosons are carrier particles that mediate the weak nuclear force, much as the photon is the carrier particle for the electromagnetic force.

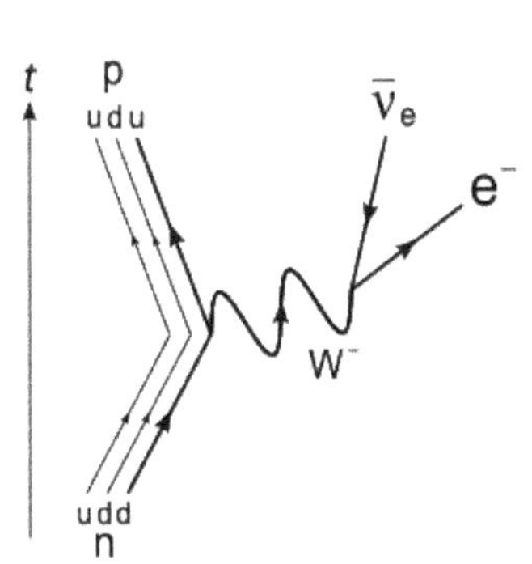

The Feynman diagram for beta decay of a neutron into a proton, electron, and electron antineutrino via an intermediate heavy W boson

W Bosons

The W bosons are best known for their role in nuclear decay. Consider, for example, the beta decay of cobalt-60.

$$^{60}_{27}\mathrm{Co} \rightarrow {}^{60}_{28}\mathrm{Ni}^{+} + e^{-} + \bar{\nu}_e$$

This reaction does not involve the whole cobalt-60 nucleus, but affects only one of its 33 neutrons. The neutron is converted into a proton while also emitting an electron (called a beta particle in this context) and an electron antineutrino:

$$n^{0} \rightarrow p^{+} + e^{-} + \bar{\nu}_e$$

Again, the neutron is not an elementary particle but a composite of an up quark and two down quarks (udd). It is in fact one of the down quarks that interacts in beta decay, turning into an up quark to form a proton (uud). At the most fundamental level, then, the weak force changes the flavour of a single quark:

$$d \rightarrow u + W^{-}$$

which is immediately followed by decay of the W− itself:

$$W^{-} \rightarrow e^{-} + \bar{\nu}_e$$

Z Boson

The Z boson is its own antiparticle. Thus, all of its flavour quantum numbers and charges are zero. The exchange of a Z boson between particles, called a neutral current interaction, therefore leaves the interacting particles unaffected, except for a transfer of momentum. Z boson interactions involving neutrinos have distinctive signatures: They provide the only known mechanism for elastic scattering of neutrinos in matter; neutrinos are almost as likely to scatter elastically (via Z boson exchange) as inelastically (via W boson exchange). The first prediction of Z bosons was made by Brazilian physicist José Leite Lopes in 1958, by devising an equation which showed the analogy of the weak nuclear interactions with electromagnetism. Steve Weinberg, Sheldon Glashow and Abdus Salam later used these results to develop the electroweak unification, in 1973. Weak neutral currents via Z boson exchange were confirmed shortly thereafter in 1974, in a neutrino experiment in the Gargamelle bubble chamber at CERN.

Predicting the W and Z

Following the spectacular success of quantum electrodynamics in the 1950s, attempts were undertaken to formulate a similar theory of the weak nuclear force. This culminated around 1968 in a unified theory of electromagnetism and weak interactions by Sheldon Glashow, Steven Weinberg, and Abdus Salam, for which they shared the 1979 Nobel Prize in Physics. Their electroweak theory postulated not only the W bosons necessary to explain beta decay, but also a new Z boson that had never been observed.

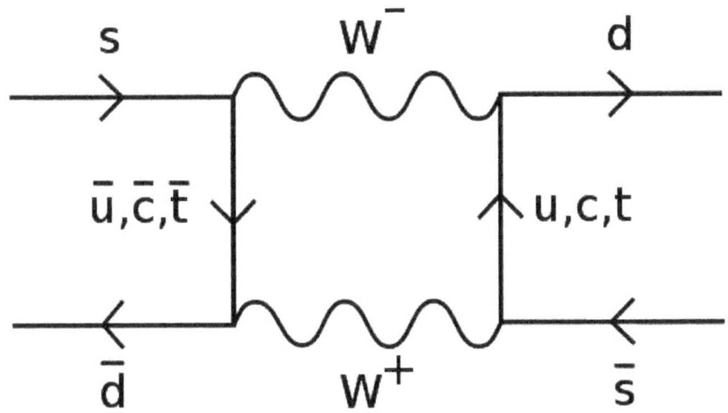

A Feynman diagram showing the exchange of a pair of W bosons. This is one of the leading terms contributing to neutral Kaon oscillation.

The fact that the W and Z bosons have mass while photons are massless was a major obstacle in developing electroweak theory. These particles are accurately described by an SU(2) gauge theory, but the bosons in a gauge theory must be massless. As a case in point, the photon is massless because electromagnetism is described by a U(1) gauge theory. Some mechanism is required to break the SU(2) symmetry, giving mass to the W and Z in the process. One explanation, the Higgs mechanism, was forwarded by the 1964 PRL symmetry breaking papers. It predicts the existence of yet another new particle; the Higgs boson. Of the four components of a Goldstone boson created by the Higgs field, three are "eaten" by the W^+, Z^0, and W^- bosons to form their longitudinal components and the remainder appears as the spin 0 Higgs boson.

The combination of the SU(2) gauge theory of the weak interaction, the electromagnetic interaction, and the Higgs mechanism is known as the Glashow-Weinberg-Salam model. These days it is widely accepted as one of the pillars of the Standard Model of particle physics. As of 13 December 2011, intensive search for the Higgs boson carried out at CERN has indicated that if the particle is to be found, it seems likely to be found around 125 GeV. On 4 July 2012, the CMS and the ATLAS experimental collaborations at CERN announced the discovery of a new particle with a mass of 125.3 ± 0.6 GeV that appears consistent with a Higgs boson.

Discovery

Unlike beta decay, the observation of neutral current interactions that involve particles other than neutrinos requires huge investments in particle accelerators and detectors, such as are available in only a few high-energy physics laboratories in the world (and then only after 1983). This is because Z-bosons behave in somewhat the same manner as photons, but do not become important until the energy of the interaction is comparable with the relatively huge mass of the Z boson.

The Gargamelle bubble chamber, now exhibited at CERN

The discovery of the W and Z bosons was considered a major success for CERN. First, in 1973, came the observation of neutral current interactions as predicted by electroweak theory. The huge Gargamelle bubble chamber photographed the tracks of a few electrons suddenly starting to move, seemingly of their own accord. This is interpreted as a neutrino interacting with the electron by the exchange of an unseen Z boson. The neutrino is otherwise undetectable, so the only observable effect is the momentum imparted to the electron by the interaction.

The discovery of the W and Z bosons themselves had to wait for the construction of a particle accelerator powerful enough to produce them. The first such machine that became available was the Super Proton Synchrotron, where unambiguous signals of W bosons were seen in January 1983 during a series of experiments made possible by Carlo Rubbia and Simon van der Meer. The actual experiments were called UA1 (led by Rubbia) and UA2 (led by Pierre Darriulat), and were the collaborative effort of many people. Van der Meer was the driving force on the accelerator end (stochastic cooling). UA1 and UA2 found the Z boson a few months later, in May 1983. Rubbia and van der Meer were promptly awarded the 1984 Nobel Prize in Physics, a most unusual step for the conservative Nobel Foundation.

The W^+, W^-, and Z^0 bosons, together with the photon (γ), comprise the four gauge bosons of the electroweak interaction.

Decay

The W and Z bosons decay to fermion–antifermion pairs but neither the W nor the Z bosons can decay into the higher-mass top quark. Neglecting phase space effects and higher order corrections, simple estimates of their branching fractions can be calculated from the coupling constants.

W Bosons

W bosons can decay to a lepton and neutrino or to an up-type quark and a down-type quark. The decay width of the W boson to a quark–antiquark pair is proportional to the corresponding squared CKM matrix element and the number of quark colors, $N_C = 3$. The decay widths for the W bosons are then proportional to:

Leptons		Up quarks		Charm quarks	
$e+$ ν_e	1	u d	$3\|V_{ud}\|^2$	c d	$3\|V_{cd}\|^2$
$\mu+$ ν_μ	1	u s	$3\|V_{us}\|^2$	c s	$3\|V_{cs}\|^2$
$\tau+$ ν_τ	1	u b	$3\|V_{ub}\|^2$	c b	$3\|V_{cb}\|^2$

Here, e^+, μ^+, τ^+ denote the three flavours of leptons (more exactly, the positive charged antileptons). ν_e, ν_μ, ν_τ denote the three flavours of neutrinos. The other particles, starting with u and $\bar{d}$, all denote quarks and antiquarks (factor N_C is applied). The various V_{ij} denote the corresponding CKM matrix coefficients.

Unitarity of the CKM matrix implies that $|V_{ud}|^2 + |V_{us}|^2 + |V_{ub}|^2 = |V_{cd}|^2 + |V_{cs}|^2 + |V_{cb}|^2 = 1$. Therefore, the leptonic branching ratios of the W boson are approximately $B(e^+\nu_e) = B(\mu^+\nu_\mu) = B(\tau^+\nu_\tau) = \frac{1}{9}$. The hadronic branching ratio is dominated by the CKM-favored ud and cs final states. The sum of the hadronic branching ratios has been measured experimentally to be 67.60±0.27%, with $B(l^+\nu_l) = 10.80\pm0.09\%$.

Z Bosons

Z bosons decay into a fermion and its antiparticle. As the Z-boson is a mixture of the pre-symmetry-breaking W^0 and B^0 bosons, each vertex factor includes a factor $T_3 - Q\sin^2\theta_W$; where T_3 is the third component of the weak isospin of the fermion, Q is the electric charge of the fermion (in units of the elementary charge), and θ_W is the weak mixing angle. Because the weak isospin is different for fermions of different chirality, either left-handed or right-handed, the coupling is different as well.

The relative strengths of each coupling can be estimated by considering that the decay rates include the square of these factors, and all possible diagrams (e.g. sum over quark families, and left and right contributions). This is just an estimate, as we are considering only tree-level diagrams in the Fermi theory.

Particles		**Effective charge**		**Relative factor**	**Branching ratio**	
Name	**Symbols**	**L**	**R**		**Predicted for $x = 0.23$**	**Experimental measurements**
Neutrinos (all)	ν_e, ν_μ, ν_τ	$\frac{1}{2}$	0	$3(\frac{1}{2})^2$	20.5%	20.00±0.06%
Charged leptons (all)	e^-, μ^-, τ^-			$3((-\frac{1}{2} + x)^2 + x^2)$	10.2%	10.097±0.003%
Electron	e^-	$-\frac{1}{2} + x$	x	$(-\frac{1}{2} + x)^2 + x^2$	3.4%	3.363±0.004%
Muon	μ^-	$-\frac{1}{2} + x$	x	$(-\frac{1}{2} + x)^2 + x^2$	3.4%	3.366±0.007%
Tau	τ^-	$-\frac{1}{2} + x$	x	$(-\frac{1}{2} + x)^2 + x^2$	3.4%	3.367±0.008%
Hadrons (all)					69.2%	69.91±0.06%
Down-type quarks	d, s, b	$-\frac{1}{2} + \frac{1}{3}x$	$\frac{1}{3}x$	$3(-\frac{1}{2} + \frac{1}{3}x)^2 + 3(\frac{1}{3}x)^2$	15.2%	15.6±0.4%
Up-type quarks	u, c	$\frac{1}{2} - \frac{2}{3}x$	$-\frac{2}{3}x$	$3(\frac{1}{2} - \frac{2}{3}x)^2 + 3(-\frac{2}{3}x)^2$	11.8%	11.6±0.6%

Here, L and R denote either the left- or right-handed chirality of the fermions respectively. (The right-handed neutrinos do not exist in the standard model. However, in some extensions beyond the standard model they do.) The notation $x = \sin^2\theta_W$ is used.

The W bosons are named after the ***w***eak force. The physicist Steven Weinberg named the additional particle the "Z particle", and later gave the explanation that it was the last additional particle needed by the model. The W bosons had already been named, and the Z bosons have ***z***ero electric charge.

The two W bosons are verified mediators of neutrino absorption and emission. During these processes, the W boson charge induces electron or positron emission or absorption, thus causing nuclear transmutation. The Z boson is not involved in the absorption or emission of electrons and positrons.

The Z boson mediates the transfer of momentum, spin and energy when neutrinos scatter elastically from matter (a process which conserves charge). Such behavior is almost as common as inelastic neutrino interactions and may be observed in bubble chambers upon irradiation with neutrino beams. Whenever an electron is observed as a new free particle suddenly moving with kinetic energy, it is inferred to be a result of a neutrino interacting directly with the electron, since this behavior happens more often when the neutrino beam is present. In this process, the neutrino simply strikes the electron and then scatters away from it, transferring some of the neutrino's momentum to the electron. Because neutrinos are neither affected by the strong force nor the electromagnetic force, and because the gravitational force between subatomic particles is negligible, such an interaction can only happen via the weak force. Since such an electron is not created from a nucleon, and is unchanged except for the new force impulse imparted by the neutrino, this weak force interaction between the neutrino and the electron must be mediated by an electromagnetically neutral, weak-force boson particle. Thus, this interaction requires a Z boson.

SCALAR BOSON

A scalar boson is a boson whose spin equals zero. Boson means that it has an integer-valued spin; the scalar fixes this value to 0.

The name "scalar boson" arises from quantum field theory. It refers to the particular transformation properties under Lorentz transformation.

Examples

- The only fundamental scalar boson in the standard model of elementary particle physics is the Higgs boson, the existence of which was confirmed on 14 March 2013 at the Large Hadron Collider. As a result of this confirmation, the 2013 Nobel Prize in physics was awarded to Peter Higgs and François Englert.
- Various known composite particles are scalar bosons, e.g. the alpha particle and the pi meson.
- Scalar mesons have invariant parity under Lorentz transformation, whereas pseudoscalar mesons reverse parity under Lorentz transformation.
- The φ^4-theory, a popular quantum field theory, uses scalar bosonic fields as employed in many introductory books to quantum field theories, serving as a toy model, introducing the basic concepts of the field.

HIGGS BOSON

The Higgs boson is an elementary particle in the Standard Model of particle physics. It is the quantum excitation of the Higgs field, a fundamental field of crucial importance to particle physics theory first suspected to exist in the 1960s. Unlike other known fields such as the electromagnetic field, it has a non-zero constant value in vacuum. The question of the Higgs field's existence has been the last unverified part of the Standard Model of particle physics and, according to some, "the central problem in particle physics".

The presence of this field, now believed to be confirmed, explains why some fundamental particles have mass when, based on the symmetries controlling their interactions, they should be massless. The existence of the Higgs field would also resolve several other long-standing puzzles, such as the reason for the weak force's extremely short range.

Although it is hypothesized that the Higgs field permeates the entire Universe, evidence for its existence has been very difficult to obtain. In principle, the Higgs field can be detected through its excitations, manifested as Higgs particles, but these are extremely difficult to produce and detect. The importance of this fundamental question led to a 40 year search, and the construction of one of the world's most expensive and complex experimental facilities to date, CERN's Large Hadron Collider, in an attempt to create Higgs bosons and other particles for observation and study. On 4 July 2012, the discovery of a new particle with a mass between 125 and 127 GeV/c^2 was announced; physicists suspected that it was the Higgs boson. Since then, the particle has been shown to behave, interact, and decay in many of the ways predicted by the Standard Model. It was also tentatively confirmed to have even parity and zero spin, two fundamental attributes of a Higgs boson. This appears to be the first elementary scalar particle discovered in nature. More studies are needed to verify that the discovered particle has properties matching those predicted for the Higgs boson by the Standard Model, or whether, as predicted by some theories, multiple Higgs bosons exist.

The Higgs boson is named after Peter Higgs, one of six physicists who, in the 1964 PRL symmetry breaking papers, proposed the Higgs mechanism that suggested the existence of such a particle. On December 10, 2013, two of them, Peter Higgs and François Englert, were awarded the Nobel Prize in Physics for their work and prediction (Englert's co-researcher Robert Brout had died in 2011 and the Nobel Prize is not ordinarily given posthumously). Although Higgs's name has come to be associated with this theory, several researchers between about 1960 and 1972 independently developed different parts of it. In mainstream media the Higgs boson has often been called the "God particle", from a 1993 book on the topic; the nickname is strongly disliked by many physicists, including Higgs, who regard it as sensationalistic.

In the Standard Model, the Higgs particle is a boson with no spin, electric charge, or colour charge. It is also very unstable, decaying into other particles almost immediately. It is a quantum excitation of one of the four components of the Higgs field. The latter constitutes a scalar field, with two neutral and two electrically charged components that form a complex doublet of the weak isospin SU(2) symmetry. The Higgs field is tachyonic—does not refer to faster-than-light speeds, but means that symmetry-breaking through condensation of a particle must occur under certain conditions. The Higgs field has a "Mexican hat" shaped potential with nonzero strength everywhere (including otherwise empty space), which in its vacuum state breaks the weak isospin symmetry of the electroweak interaction. When this happens, three components of the Higgs field are "absorbed" by the SU(2) and U(1) gauge bosons (the "Higgs mechanism") to become the longitudinal components of the now-massive W and Z bosons of the weak force. The remaining electrically neutral component separately couples to other particles known as fermions (via Yukawa couplings), causing these to acquire mass as well. Some versions of the theory predict more than one kind of Higgs fields and bosons. Alternative "Higgsless" models may have been considered if the Higgs boson had not been discovered.

Non-Technical Summary

The Standard Model

Physicists explain the properties and forces between elementary particles in terms of the Standard Model—a widely accepted and "remarkably" accurate framework for understanding almost everything in the known universe, other than gravity. In the model, the fundamental forces of nature arise from gauge invariance and symmetries. Forces are transmitted by particles known as gauge bosons.

The Problem of Gauge Boson Mass

By around 1960 all attempts to create a gauge invariant theory for the fundamental forces of electromagnetism and the weak force had consistently failed. The problem was that the symmetry requirements in gauge theory predicted that both electromagnetism's gauge boson (the photon) and the weak force's gauge bosons (W and Z) should have zero mass.

However, although the photon is indeed massless, experiments show that these bosons have mass.[:372-373]Explaining this unexpected mass was a significant problem in the Standard Model.

Symmetry Breaking

By the early 1960s, physicists had realised that a given symmetry law might not always be followed under certain conditions. In 1962 physicist Philip Anderson wrote a paper that built upon work by Yoichiro Nambu concerning "broken symmetries" in superconductivity and particle physics. He suggested that "broken symmetries" might also be the missing piece needed to solve the problems of gauge invariance. It was hoped that symmetry breaking might explain why electromagnetism's boson is massless, yet the weak force bosons have mass.

Higgs Mechanism

The Higgs mechanism was proposed in 1964 by three groups of researchers as a possible explanation for why gauge bosons could still be massive despite their governing symmetry. It postulated that the conditions for the symmetry would be 'broken' if an unusual type of field existed throughout the Universe, and then why some fundamental particles would be able to acquire mass.

Although these ideas did not gain much initial support or attention, by 1972 they had been developed into a comprehensive theory and proved capable of giving "sensible" results that accurately described particles known at the time, and which accurately predicted several other particles discovered during the following years. During the 1970s these theories rapidly became the "standard model". There was not yet any direct evidence that the Higgs field actually existed, but even without proof of the field, the accuracy of its predictions led scientists to believe the theory might be true. By the 1980s the question whether or not the Higgs field existed had come to be regarded as one of the most important unanswered questions in particle physics.

Higgs Field

According to the Standard Model, a field of the necessary kind (the "Higgs field") exists throughout space, and breaks certain symmetry laws of the electroweak interaction. The existence of this field triggers the Higgs mechanism, causing the gauge bosons responsible for the weak force to be massive, and explaining their very short range. Furthermore, it was later realised that the same field would also explain, in a different way, why other fundamental constituents of matter (including electrons and quarks) have mass.

For many decades, scientists had no way to tell whether or not the Higgs field existed. If it did, then it would be unlike other fundamental fields. But it was also possible that these key ideas, or even the entire Standard Model itself, were somehow incorrect. Only discovering what was breaking this symmetry would solve the problem.

Higgs Boson

The hypothesised Higgs mechanism made accurate predictions,[22] however to confirm it there was an extensive search for a matching particle associated with it — the "Higgs boson". Detecting Higgs bosons was experimentally difficult due to the energy required to produce them, and that they break apart in about a ten-sextillionth (10^{-22}) of a second. It was therefore several decades before the first evidence of the Higgs boson began to be found. Particle colliders, detectors, and computers capable of looking for Higgs bosons took over 30 years (c. 1980–2010) to develop.

By March 2016, the existence of the Higgs boson was confirmed, and therefore the concept of some type of Higgs field throughout space is strongly supported. The nature and properties of this field are now being investigated. Discovery of the Higgs boson, and therefore support of the Higgs field was a monumental discovery for science and human knowledge, and opened doorways to new knowledge in many disciplines.

Interpretation

Various analogies have been used to describe the Higgs field and boson, including analogies with well-known symmetry breaking effects such as the rainbow and prism, electric fields, ripples, and resistance of macro objects moving through media, like people moving through crowds or some objects moving through syrup or molasses. However, analogies based on simple resistance to motion are inaccurate as the Higgs field does not work by resisting motion.

SIGNIFICANCE

Evidence of the Higgs field and its properties has been extremely significant scientifically, for many reasons. The Higgs boson's importance is largely that it is able to be examined using existing knowledge and experimental technology, as a way to confirm and study the entire Higgs field theory. Conversely, proof that the Higgs field and boson do not exist would also have been significant.

Particle Physics

Validation of the Standard Model

The Higgs boson may validate the Standard Model through the mechanism of mass generation. If it does, then more advanced extensions are can be suggested or excluded based upon measurements of its properties. As experimental means to measure the field's behaviours and interactions are developed, this fundamental field can be better understood. Alternatively, if the Higgs field doesn't exist, the Standard Model may need to be extended or superseded (such as supersymmetry or string theory).

Related to this, a belief generally exists among physicists that there is likely to be "new" physics beyond the Standard Model—the Standard Model will at some point be extended or superseded. The Higgs field and related issues present a promising "doorway" to understand better the places where the Standard Model might become inadequate or fail, and could provide considerable evidence guiding researchers into future enhancements or successors.

Symmetry Breaking of the Electroweak Interaction

Below an extremely high temperature, electroweak symmetry breaking causes the electroweak interaction to manifest in part as the short-ranged weak force, which is carried by massive gauge bosons. This symmetry breaking is required for atoms and other structures to form, as well as nuclear reactions in stars such as our Sun. However, it is not clear how symmetry breaking actually happens in nature. The Higgs field may provide a mechanism for this observed symmetry breaking. Experimental measurements will be needed to validate whether the Standard Model is correct in its approach, and if it can be made more exact.

Particle Mass Acquisition

Electroweak symmetry breaking (due to a Higgs field or otherwise) is believed proven responsible for generating the masses of fundamental particles such as elementary fermions (including electrons and quarks) and the massive W and Z gauge bosons. Finding how this happens is pivotal to particle physics.

It is worth noting that the Higgs field does not "create" mass out of nothing (which would violate the law of conservation of energy), nor is the Higgs field responsible for the mass of all particles. For example, about 99% of the mass of baryons (composite particles such as the proton and neutron) is due instead to quantum chromodynamics binding energy, which is the sum of the kinetic energies of quarks and the energies of the massless gluons mediating the strong interaction inside the baryons. In Higgs-based theories, the property of "mass" is a manifestation of potential energy transferred to particles when they interact ("couple") with the Higgs field, which had contained that mass in the form of energy.

Scalar Fields and Extension of the Standard Model

The significance of proof of a scalar field such as the Higgs field would be hard to overestimate: "[The] verification of real scalar fields would be nearly as important as its role in generating mass". Rolf-Dieter Heuer, director general of CERN, stated in a 2011 talk on the Higgs field:

All the matter particles are spin-1/2 fermions. All the force carriers are spin-1 bosons. Higgs particles are spin-0 bosons (scalars). The Higgs is neither matter nor force. The Higgs is just different. This would be the first fundamental scalar ever discovered. The Higgs field is thought to fill the entire universe. Could it give some handle of dark energy (scalar field)? Many modern theories predict other scalar particles like the Higgs. Why, after all, should the Higgs be the only one of its kind? [The] LHC can search for and study new scalars with precision.

Cosmology

Inflation

There has been considerable scientific research on possible links between the Higgs field and the inflaton – a hypothetical field suggested as the explanation for the expansion of space during the first fraction of a second of the universe (known as the "inflationary epoch"). Some theories suggest that a fundamental scalar field might be responsible for this phenomenon; the Higgs field is such a field and therefore has led to papers analysing whether it could also be the inflaton responsible for this exponential expansion of the universe during the Big Bang. Such theories are highly tentative and face significant problems related to unitarity, but may be viable if combined with additional features such as large non-minimal coupling, a Brans–Dicke scalar, or other "new" physics, and have received treatments suggesting that Higgs inflation models are still of interest theoretically.

Nature and Fate of the Universe, and its Possible Fates

For decades, scientific models of our universe have included the possibility that it exists as a long-lived, but not completely stable, sector of space, which could potentially at some time be destroyed upon 'toppling' into a more stable vacuum state. If the masses of the Higgs boson and top quark are known more exactly, and the Standard Model provides an accurate description of particle physics up to extreme energies of the Planck scale, then it is possible to calculate whether the universe's present vacuum state is stable or merely long-lived. (This was sometimes misreported as the Higgs boson "ending" the universe).

A 125 – 127 GeV Higgs mass seems to be extremely close to the boundary for stability (estimated in 2012 as 123.8 – 135.0 GeV) but a definitive answer requires much more precise measurements of the top quark's pole mass. New physics can change this picture.

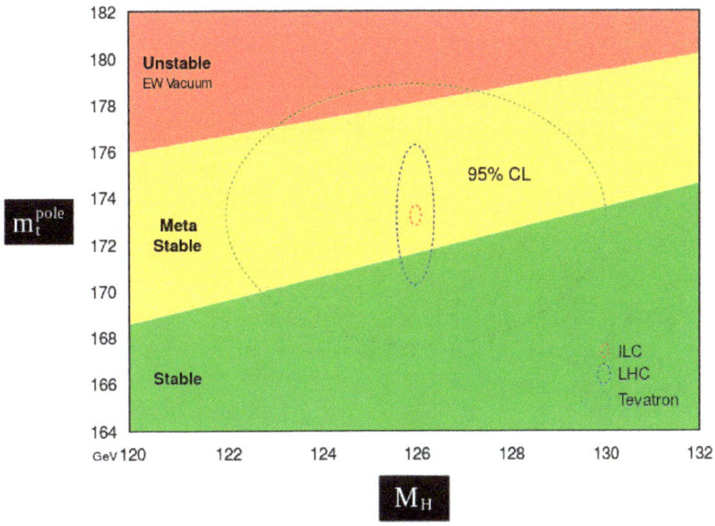

Diagram showing the Higgs boson and top quarkmasses, which could indicate whether our universe is stable, or a long-lived 'bubble'. As of 2012, the 2σ ellipse based onTevatron and LHC data still allows for both possibilities.

If measurements of the Higgs boson suggest that our universe lies within a false vacuum of this kind, then it would imply – more than likely in many billions of years – that the universe's forces, particles, and structures could cease to exist as we know them (and be replaced by different ones), if a true vacuum happened to nucleate. It also suggests that the Higgs self-coupling λ and its β_λ function could be very close to zero at the Planck scale, with «intriguing» implications, including theories of gravity and Higgs-based inflation.[218] A future electron–positron collider would be able to provide the precise measurements of the top quark needed for such calculations.

Vacuum Energy and the Cosmological Constant

More speculatively, the Higgs field has also been proposed as the energy of the vacuum, which at the extreme energies of the first moments of the Big Bang caused the universe to be a kind of featureless symmetry of undifferentiated extremely high energy. In this kind of speculation, the single unified field of a Grand Unified Theory is identified as (or modelled upon) the Higgs field, and it is through successive symmetry breakings of the Higgs field or some similar field at phase transitions that the present universe's known forces and fields arise.

The relationship (if any) between the Higgs field and the presently observed vacuum energy density of the universe has also come under scientific study. As observed, the present vacuum energy density is extremely close to zero, but the energy density expected from the Higgs field, supersymmetry, and other current theories are typically many orders of magnitude larger. It is unclear how these should be reconciled. This cosmological constant problem remains a further major unanswered problem in physics.

Practical and Technological Impact

As yet, there are no known immediate technological benefits of finding the Higgs particle. However, a common pattern for fundamental discoveries is for practical applications to follow later, once the discovery has been explored further, at which point they become the basis for new technologies of importance to society.

The challenges in particle physics have furthered major technological progress of widespread importance. For example, the World Wide Web began as a project to improve CERN's communication system. CERN's requirement to process massive amounts of data produced by the Large Hadron Collider also led to contributions to the fields of distributed and cloud computing.

History

Particle physicists study matter made from fundamental particles whose interactions are mediated by exchange particles – gauge bosons – acting as force carriers. At the beginning of the 1960s a number of these particles had been discovered or proposed, along with theories suggesting how they relate to each other, some of which had already been reformulated as field theories in which the objects of study are not particles and forces, but quantum fields and their symmetries.[:150] However, attempts to unify known fundamental forces such as the electromagnetic force and the weak nuclear force were known to be incomplete. One known omission was that gauge invariant approaches, including non-abelian models such as Yang–Mills theory (1954), which held great promise for unified theories, also seemed to predict known massive particles as massless. Goldstone's theorem, relating to continuous symmetries within some theories, also appeared to rule out many obvious solutions, since it appeared to show that zero-mass particles would have to also exist that were "simply not seen".According to Guralnik, physicists had "no understanding" how these problems could be overcome.

Yang and Mills work on non-abelian gauge theory had one huge problem: in perturbation theory it has massless particles which don't correspond to anything we see. One way of getting rid of this problem is now fairly well-understood, the phenomenon of confinement realized in QCD, where the strong interactions get rid of the massless "gluon" states at long distances. By the very early sixties, people had begun to understand another source of massless particles: spontaneous symmetry breaking of a continuous symmetry. What Philip Anderson realized and worked out in the summer of 1962 was that, when you have both gauge symmetry and spontaneous symmetry breaking, the Nambu–Goldstone massless mode can combine with the massless gauge field modes to produce a physical massive vector field. This is what happens in superconductivity, a subject about which Anderson was (and is) one of the leading experts.

Nobel Prize Laureate Peter Higgs in Stockholm, December 2013

Particle physicist and mathematician Peter Woit summarised the state of research at the time:

The Higgs mechanism is a process by which vector bosons can get rest mass without explicitly breaking gauge invariance, as a byproduct of spontaneous symmetry breaking. The mathematical theory behind spontaneous symmetry breaking was initially conceived and published within particle physics by Yoichiro Nambu in 1960, the concept that such a mechanism could offer a possible solution for the "mass problem" was originally suggested in 1962 by Philip Anderson (who had previously written papers on broken symmetry and its outcomes in superconductivity and concluded in his 1963 paper on Yang-Mills theory that "considering the superconducting analog... [t]hese two types of bosons seem capable of canceling each other out... leaving finite mass bosons"),[4–5] and Abraham Klein and Benjamin Lee showed in March 1964 that Goldstone's theorem could be avoided this way in at least some non-relativistic cases and speculated it might be possible in truly relativistic cases.

These approaches were quickly developed into a full relativistic model, independently and almost simultaneously, by three groups of physicists: by François Englert and Robert Brout in August 1964; by Peter Higgs in October 1964; and by Gerald Guralnik, Carl Hagen, and Tom Kibble (GHK) in November 1964. Higgs also wrote a short but important response published in September 1964 to an objection by Gilbert, which showed that if calculating within the radiation gauge, Goldstone's theorem and Gilbert's objection would become inapplicable. (Higgs later described Gilbert's objection as prompting his own paper.) Properties of the model were further considered by Guralnik in 1965, by Higgs in 1966, by Kibble in 1967, and further by GHK in 1967. The original three 1964 papers showed that when a gauge theory is combined with an additional field that spontaneously breaks the symmetry, the gauge bosons can consistently acquire a finite mass. In 1967, Steven Weinberg and Abdus Salam independently showed how a Higgs mechanism could be used to break the electroweak symmetry of Sheldon Glashow's unified model for the weak and electromagnetic interactions (itself an extension of work by Schwinger), forming what became the Standard Model of particle physics. Weinberg was the first to observe that this would also provide mass terms for the fermions. However, the seminal papers on spontaneous breaking of gauge symmetries were at first largely ignored, because it was widely believed that the (non-Abelian gauge) theories in question were a dead-end, and in particular that they could not be renormalised. In 1971–72, Martinus Veltman and Gerard 't Hooft proved renormalisation of Yang–Mills was possible in two papers covering massless, and then massive, fields. Their contribution, and others' work on the renormalisation group – including "substantial" theoretical work by Russian physicists Ludvig Faddeev, Andrei Slavnov, Efim Fradkin and Igor Tyutin – was eventually "enormously profound and influential", but even with all key elements of the eventual theory published there was still almost no wider interest. For example, Coleman found in a study that "essentially no-one paid any attention" to Weinberg's paper prior to 1971 and discussed by David Politzer in his 2004 Nobel speech. – now the most cited in particle physics – and even in 1970 according to Politzer, Glashow's teaching of the weak interaction contained no mention of Weinberg's, Salam's, or Glashow's own work. In practice, Politzer states, almost everyone learned of the theory due to physicist Benjamin Lee, who combined the work of Veltman and 't Hooft with insights by others, and popularised the completed theory. In this way, from 1971, interest and acceptance "exploded" and the ideas were quickly absorbed in the mainstream.

The resulting electroweak theory and Standard Model have accurately predicted (among other things) weak neutral currents, three bosons, the top and charm quarks, and with great precision, the mass and other properties of some of these. Many of those involved eventually won Nobel Prizes or other renowned awards. A 1974 paper and comprehensive review in Reviews of Modern Physics commented that "while no one doubted the [mathematical] correctness of these arguments, no one quite believed that nature was diabolically clever enough to take advantage of them",[9] adding that the theory had so far produced accurate answers that accorded with experiment, but it was unknown whether the theory was fundamentally correct. By 1986 and again in the 1990s it became possible to write that understanding and proving the Higgs sector of the Standard Model was "the central problem today in particle physics".

Summary and Impact of the PRL Papers

The three papers written in 1964 were each recognised as milestone papers during Physical Review Letters's 50th anniversary celebration. Their six authors were also awarded the 2010 J. J. Sakurai Prize for Theoretical Particle Physics for this work. (A controversy also arose the same year, because in the event of a Nobel Prize only up to three scientists could be recognised, with six being credited for the papers.) Two of the three PRL papers (by Higgs and by GHK) contained equations for the hypothetical field that eventually would become known as the Higgs field and its hypothetical quantum, the Higgs boson. Higgs' subsequent 1966 paper showed the decay mechanism of the boson; only a massive boson can decay and the decays can prove the mechanism.

In the paper by Higgs the boson is massive, and in a closing sentence Higgs writes that "an essential feature" of the theory "is the prediction of incomplete multiplets of scalar and vector bosons". (Frank Close comments that 1960s gauge theorists were focused on the problem of massless vector bosons, and the implied existence of a massive scalar boson was not seen as important; only Higgs directly addressed it. In the paper by GHK the boson is massless and decoupled from the massive states. In reviews dated 2009 and 2011, Guralnik states that in the GHK model the boson is massless only in a lowest-order approximation, but it is not subject to any constraint and acquires mass at higher orders, and adds that the GHK paper was the only one to show that there are no massless Goldstone bosons in the model and to give a complete analysis of the general Higgs mechanism. All three reached similar conclusions, despite their very different approaches: Higgs' paper essentially used classical techniques, Englert and Brout's involved calculating vacuum polarisation in perturbation theory around an assumed symmetry-breaking vacuum state, and GHK used operator formalism and conservation laws to explore in depth the ways in which Goldstone's theorem may be worked around.

Theoretical Properties

Theoretical Need for the Higgs

Gauge invariance is an important property of modern particle theories such as the Standard Model, partly due to its success in other areas of fundamental physics such as electromagnetism and the strong interaction (quantum chromodynamics). However, there were great difficulties in developing gauge theories for the weak nuclear force or a possible unified electroweak interaction. Fermions with a mass term would violate gauge symmetry and therefore cannot be gauge invariant. (This can be seen by examining the Dirac Lagrangian for a fermion in terms of left and right handed components; we find none of the spin-half particles could ever flip helicity as required for mass, so they must be massless.) W and Z bosons are observed to have mass, but a boson mass term contains terms, which clearly depend on the choice of gauge and therefore these masses too cannot be gauge invariant. Therefore, it seems that none of the standard model fermions or bosons could "begin" with mass as an inbuilt property except by abandoning gauge invariance. If gauge invariance were to be retained, then these particles had to be acquiring their mass by some other mechanism or interaction. Additionally, whatever was giving these particles their mass, had to not "break" gauge invariance as the basis for other parts of the theories where it worked well, and had to not require or predict unexpected massless particles and long-range forces (seemingly an inevitable consequence of Goldstone's theorem) which did not actually seem to exist in nature.

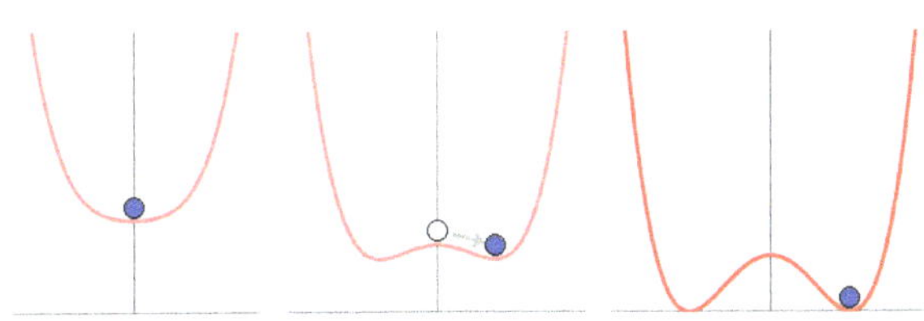

"Symmetry breaking illustrated": – At high energy levels (left) the ball settles in the centre, and the result is symmetrical. At lower energy levels (right), the overall "rules" remain symmetrical, but the "Mexican hat" potential comes into effect: "local" symmetry inevitably becomes broken since eventually the ball must at random roll one way or another.

A solution to all of these overlapping problems came from the discovery of a previously unnoticed borderline case hidden in the mathematics of Goldstone's theorem, that under certain conditions it might theoretically be possible for a symmetry to be broken without disrupting gauge invariance and without any new massless particles or forces, and having "sensible" (renormalisable) results mathematically: this became known as the Higgs mechanism.

The Standard Model hypothesises a field which is responsible for this effect, called the Higgs field (symbol: ϕ), which has the unusual property of a non-zero amplitude in its ground state; i.e., a non-zero vacuum expectation value. It can have this effect because of its unusual "Mexican hat" shaped potential whose lowest "point" is not at its "centre". Below a certain extremely high energy level the existence of this non-zero vacuum expectation spontaneously breaks electroweakgauge symmetry which in turn gives rise to the Higgs mechanism and triggers the acquisition of mass by those particles interacting with the field. This effect occurs because scalar field components of the Higgs field are "absorbed" by the massive bosons as degrees of freedom, and couple to the fermions via Yukawa coupling, thereby producing the expected mass terms. In effect when symmetry breaks under these conditions, the Goldstone bosons that arise interact with the Higgs field (and with other particles capable of interacting with the Higgs field) instead of becoming new massless particles, the intractable problems of both underlying theories "neutralise" each other, and the residual outcome is that elementary particles acquire a consistent mass based on how strongly they interact with the Higgs field. It is the simplest known process capable of giving mass to the gauge bosons while remaining compatible with gauge theories. Its quantum would be a scalar boson, known as the Higgs boson.

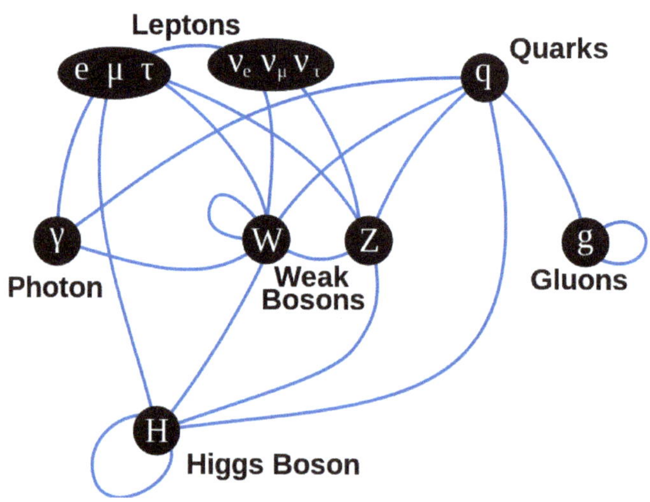

Summary of interactions between certain particles described by the Standard Model.

Properties of the Higgs field

In the Standard Model, the Higgs field is a scalar tachyonic field – scalar meaning it does not transform under Lorentz transformations, and tachyonic meaning the field (but **not** the particle) has imaginary mass, and in certain configurations must undergo symmetry breaking. It consists of four components: two neutral ones and two charged component fields. Both of the charged components and one of the neutral fields are Goldstone bosons, which act as the longitudinal third-polarisation components of the massive W^+, W^-, and Z bosons. The quantum of the remaining neutral component corresponds to (and is theoretically realised as) the massive Higgs boson, this component can interact with fermions via Yukawa coupling to give them mass, as well.

Mathematically, the Higgs field has imaginary mass and is therefore a tachyonic field. While tachyons (particles that move faster than light) are a purely hypothetical concept, fields with imaginary mass have come to play an important role in modern physics. Under no circumstances do any excitations ever propagate faster than light in such theories — the presence or absence of a tachyonic mass has no effect whatsoever on the maximum velocity of signals (there is no violation of causality). Instead of faster-than-light particles, the imaginary mass creates an instability: Any configuration in which one or more field excitations are tachyonic must spontaneously decay, and the resulting configuration contains no physical tachyons. This process is known as tachyon condensation, and is now believed to be the explanation for how the Higgs mechanism itself arises in nature, and therefore the reason behind electroweak symmetry breaking.

Although the notion of imaginary mass might seem troubling, it is only the field, and not the mass itself, that is quantised. Therefore, the field operators at spacelike separated points still commute (or anticommute), and information and particles still do not propagate faster than light. Tachyon condensation drives a physical system that has reached a local limit – and might naively be expected to produce physical tachyons – to an alternate stable state where no physical tachyons exist. Once a tachyonic field such as the Higgs field reaches the minimum of the potential, its quanta are not tachyons any more but rather are ordinary particles such as the Higgs boson.

Properties of the Higgs boson

Since the Higgs field is scalar, the Higgs boson has no spin. The Higgs boson is also its own antiparticle and is CP-even, and has zero electric and colour charge.

The Minimal Standard Model does not predict the mass of the Higgs boson. If that mass is between 115 and 180 GeV/c^2, then the Standard Model can be valid at energy scales all the way up to the Planck scale (10^{19} GeV). Many theorists expect new physics beyond the Standard Model to emerge at the TeV-scale, based on unsatisfactory properties of the Standard Model. The highest possible mass scale allowed for the Higgs boson (or some other electroweak symmetry breaking mechanism) is 1.4 TeV; beyond this point, the Standard Model becomes inconsistent without such a mechanism, because unitarity is violated in certain scattering processes.

It is also possible, although experimentally difficult, to estimate the mass of the Higgs boson indirectly. In the Standard Model, the Higgs boson has a number of indirect effects; most notably, Higgs loops result in tiny corrections to masses of W and Z bosons. Precision measurements of electroweak parameters, such as the Fermi constant and masses of W/Z bosons, can be used to calculate constraints on the mass of the Higgs. As of July 2011, the precision electroweak measurements tell us that the mass of the Higgs boson is likely to be less than about 161 GeV/c^2 at 95% confidence level (this upper limit would increase to 185 GeV/c^2 if the lower bound of 114.4 GeV/c^2 from the LEP-2 direct search is allowed for). These indirect constraints rely on the assumption that the Standard Model is correct. It may still be possible to discover a Higgs boson above these masses if it is accompanied by other particles beyond those predicted by the Standard Model.

Production

If Higgs particle theories are valid, then a Higgs particle can be produced much like other particles that are studied, in a particle collider. This involves accelerating a large number of particles to extremely high energies and extremely close to the speed of light, then allowing them to smash together. Protons and lead ions (the bare nuclei of lead atoms) are used at the LHC. In the extreme energies of these collisions, the desired esoteric particles will occasionally be produced and this can be detected and studied; any absence or difference from theoretical expectations can also be used to improve the theory.

Feynman diagrams for Higgs production

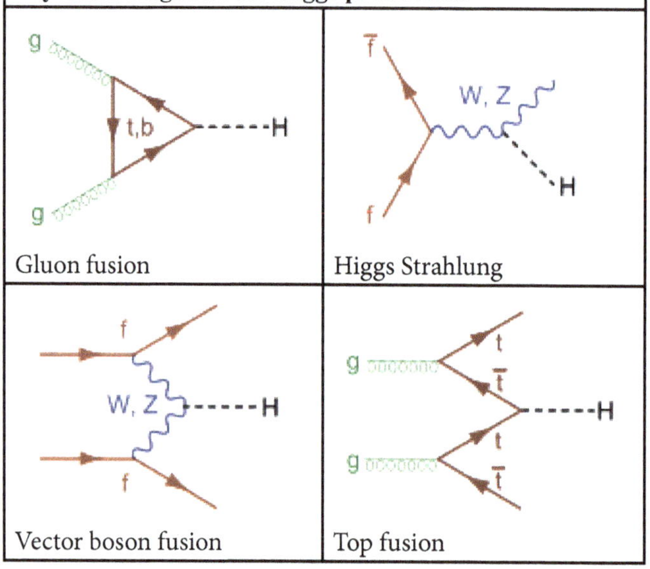

The relevant particle theory (in this case the Standard Model) will determine the necessary kinds of collisions and detectors. The Standard Model predicts that Higgs bosons could be formed in a number of ways, although the probability of producing a Higgs boson in any collision is always expected to be very small—for example, only 1 Higgs boson per 10 billion collisions in the Large Hadron Collider. The most common expected processes for Higgs boson production are:

- Gluon fusion. If the collided particles are hadrons such as the proton or antiproton—as is the case in the LHC and Tevatron—then it is most likely that two of the gluons binding the hadron together collide. The easiest way to produce a Higgs particle is if the two gluons combine to form a loop of virtual quarks. Since the coupling of particles to the Higgs boson is proportional to their mass, this process is more likely for heavy particles. In practice it is enough to consider the contributions of virtual top and bottom quarks (the heaviest quarks). This process is the dominant contribution at the LHC and Tevatron being about ten times more likely than any of the other processes.

- Higgs Strahlung. If an elementary fermion collides with an anti-fermion—e.g., a quark with an anti-quark or an electron with a positron—the two can merge to form a virtual W or Z boson which, if it carries sufficient energy, can then emit a Higgs boson. This process was the dominant production mode at the LEP, where an electron and a positron collided to form a virtual Z boson, and it was the second largest contribution for Higgs production at the Tevatron. At the LHC this process is only the third largest, because the LHC collides protons with protons, making a quark-antiquark collision less likely than at the Tevatron. Higgs Strahlung is also known as associated production.
- Weak boson fusion. Another possibility when two (anti-)fermions collide is that the two exchange a virtual W or Z boson, which emits a Higgs boson. The colliding fermions do not need to be the same type. So, for example, an up quark may exchange a Z boson with an anti-down quark. This process is the second most important for the production of Higgs particle at the LHC and LEP.
- Top fusion. The final process that is commonly considered is by far the least likely (by two orders of magnitude). This process involves two colliding gluons, which each decay into a heavy quark–antiquark pair. A quark and antiquark from each pair can then combine to form a Higgs particle.

Decay

Quantum mechanics predicts that if it is possible for a particle to decay into a set of lighter particles, then it will eventually do so. This is also true for the Higgs boson.

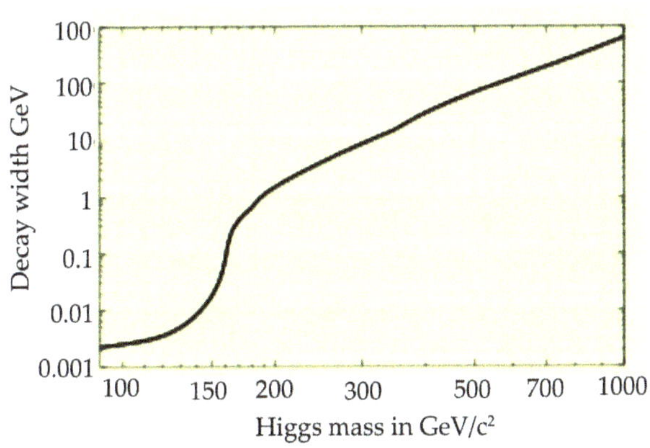

The Standard Model prediction for the decay width of the Higgs particle depends on the value of its mass.

The likelihood with which this happens depends on a variety of factors including: the difference in mass, the strength of the interactions, etc. Most of these factors are fixed by the Standard Model, except for the mass of the Higgs boson itself. For a Higgs boson with a mass of 126 GeV/c^2 the SM predicts a mean life time of about 1.6×10^{-22} s.

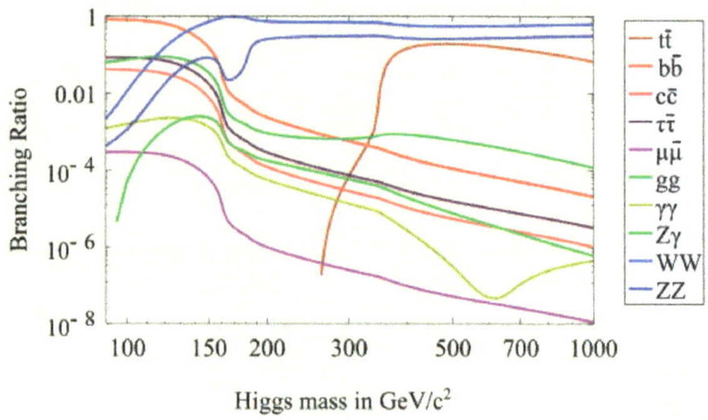

The Standard Model prediction for the branching ratios of the different decay modes of the Higgs particle depends on the value of its mass.

Since it interacts with all the massive elementary particles of the SM, the Higgs boson has many different processes through which it can decay. Each of these possible processes has its own probability, expressed as the branching ratio; the fraction of the total number decays that follows that process. The SM predicts these branching ratios as a function of the Higgs mass.

One way that the Higgs can decay is by splitting into a fermion–antifermion pair. As general rule, the Higgs is more likely to decay into heavy fermions than light fermions, because the mass of a fermion is proportional to the strength of its interaction with the Higgs. By this logic the most common decay should be into a top–antitop quark pair. However, such a decay is only possible if the Higgs is heavier than ~346 GeV/c^2, twice the mass of the top quark. For a Higgs mass of 126 GeV/c^2 the SM predicts that the most common decay is into a bottom–antibottom quark pair, which happens 56.1% of the time. The second most common fermion decay at that mass is a tau–antitau pair, which happens only about 6% of the time.

Another possibility is for the Higgs to split into a pair of massive gauge bosons. The most likely possibility is for the Higgs to decay into a pair of W bosons (the light blue line in the plot), which happens about 23.1% of the time for a Higgs boson with a mass of 126 GeV/c^2. The W bosons can subsequently decay either into a quark and an antiquark or into a charged lepton and a neutrino. However, the decays of W bosons into quarks are difficult to distinguish from the background, and the decays into leptons cannot be fully reconstructed (because neutrinos are impossible to detect in particle collision experiments). A cleaner signal is given by decay into a pair of Z-bosons (which happens about 2.9% of the time for a Higgs with a mass of 126 GeV/c^2), if each of the bosons subsequently decays into a pair of easy-to-detect charged leptons (electrons or muons).

Decay into massless gauge bosons (i.e., gluons or photons) is also possible, but requires intermediate loop of virtual heavy quarks (top or bottom) or massive gauge bosons. The most common such process is the decay into a pair of gluons through a loop of virtual heavy quarks. This process, which is the reverse of the gluon fusion process mentioned above, happens approximately 8.5% of the time for a Higgs boson with a mass of 126 GeV/c^2. Much rarer is the decay into a pair of photons mediated by a loop of W bosons or heavy quarks, which happens only twice for every thousand decays. However, this process is very relevant for experimental searches for the Higgs boson, because the energy and momentum of the photons can be measured very precisely, giving an accurate reconstruction of the mass of the decaying particle.

Alternative Models

The Minimal Standard Model as described above is the simplest known model for the Higgs mechanism with just one Higgs field. However, an extended Higgs sector with additional Higgs particle doublets or triplets is also possible, and many extensions of the Standard Model have this feature. The non-minimal Higgs sector favoured by theory are the two-Higgs-doublet models (2HDM), which predict the existence of a quintet of scalar particles: two CP-even neutral Higgs bosons h^0 and H^0, a CP-odd neutral Higgs boson A^0, and two charged Higgs particles $H^{\pm}$. Supersymmetry («SUSY») also predicts relations between the Higgs-boson masses and the masses of the gauge bosons, and could accommodate a 125 GeV/c^2 neutral Higgs boson.

The key method to distinguish between these different models involves study of the particles' interactions ("coupling") and exact decay processes ("branching ratios"), which can be measured and tested experimentally in particle collisions. In the Type-I 2HDM model one Higgs doublet couples to up and down quarks, while the second doublet does not couple to quarks. This model has two interesting limits, in which the lightest Higgs couples to just fermions ("gauge-phobic") or just gauge bosons ("fermiophobic"), but not both. In the Type-II 2HDM model, one Higgs doublet only couples to up-type quarks, the other only couples to down-type quarks. The heavily researched Minimal Supersymmetric Standard Model (MSSM) includes a Type-II 2HDM Higgs sector, so it could be disproven by evidence of a Type-I 2HDM Higgs.

In other models the Higgs scalar is a composite particle. For example, in technicolor the role of the Higgs field is played by strongly bound pairs of fermions called techniquarks. Other models, feature pairs of top quarks. In yet other models, there is no Higgs field at all and the electroweak symmetry is broken using extra dimensions.

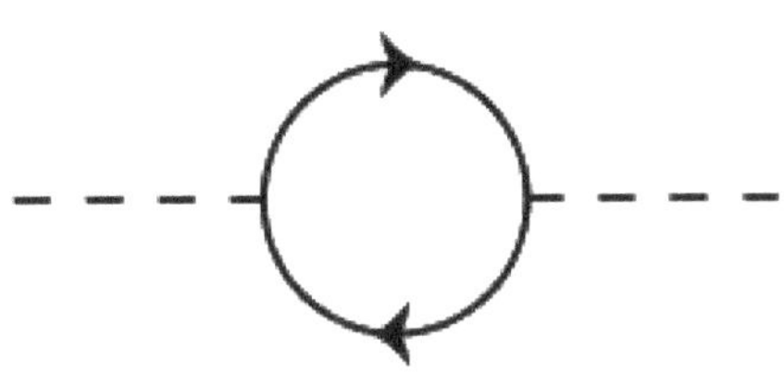

A one-loop Feynman diagram of the first-order correction to the Higgs mass. In the Standard Model the effects of these corrections are potentially enormous, giving rise to the so-called hierarchy problem.

Further Theoretical Issues and Hierarchy Problem

The Standard Model leaves the mass of the Higgs boson as a parameter to be measured, rather than a value to be calculated. This is seen as theoretically unsatisfactory, particularly as quantum corrections (related to interactions with virtual particles) should apparently cause the Higgs particle to have a mass immensely higher than that observed, but at the same time the Standard Model requires a mass of the order of 100 to 1000 GeV to ensure unitarity (in this case, to unitarise longitudinal vector boson scattering). Reconciling these points appears to require explaining why there is an almost-perfect cancellation resulting in the visible mass of ~ 125 GeV, and it is not clear how to do this. Because the weak force is about 10^{32} times stronger than gravity, and (linked to this) the Higgs boson's mass is so much less than the Planck mass or the grand unification energy, it appears that either there is some underlying connection or reason for these observations which is unknown and not described by the Standard Model, or some unexplained and extremely precise fine-tuning of parameters – however at present neither of these explanations is proven. This is known as a hierarchy problem. More broadly, the hierarchy problem amounts to the worry that a future theory of fundamental particles and interactions should not have excessive fine-tunings or unduly delicate cancellations, and should allow masses of particles such as the Higgs boson to be calculable. The problem is in some ways unique to spin-0 particles (such as the Higgs boson), which can give rise to issues related to quantum corrections that do not affect particles with spin. A number of solutions have been proposed, including supersymmetry, conformal solutions and solutions via extra dimensions such as braneworld models.

There are also issues of quantum triviality, which suggests that it may not be possible to create a consistent quantum field theory involving elementary scalar particles.

Experimental Search

To produce Higgs bosons, two beams of particles are accelerated to very high energies and allowed to collide within a particle detector. Occasionally, although rarely, a Higgs boson will be created fleetingly as part of the collision byproducts. Because the Higgs boson decays very quickly, particle detectors cannot detect it directly. Instead the detectors register all the decay products (the decay signature) and from the data the decay process is reconstructed. If the observed decay products match a possible decay process (known as a decay channel) of a Higgs boson, this indicates that a Higgs boson may have been created. In practice, many processes may produce similar decay signatures. Fortunately, the Standard Model precisely predicts the likelihood of each of these, and each known process, occurring. So, if the detector detects more decay signatures consistently matching a Higgs boson than would otherwise be expected if Higgs bosons did not exist, then this would be strong evidence that the Higgs boson exists.

Because Higgs boson production in a particle collision is likely to be very rare (1 in 10 billion at the LHC), and many other possible collision events can have similar decay signatures, the data of hundreds of trillions of collisions needs to be analysed and must "show the same picture" before a conclusion about the existence of the Higgs boson can be reached. To conclude that a new particle has been found, particle physicists require that the statistical analysis of two independent particle detectors each indicate that there is lesser than a one-in-a-million chance that the observed decay signatures are due to just background random Standard Model events—i.e., that the observed number of events is more than 5 standard deviations (sigma) different from that expected if there was no new particle. More collision data allows better confirmation of the physical properties of any new particle observed, and allows physicists to decide whether it is indeed a Higgs boson as described by the Standard Model or some other hypothetical new particle.

To find the Higgs boson, a powerful particle accelerator was needed, because Higgs bosons might not be seen in lower-energy experiments. The collider needed to have a high luminosity in order to ensure enough collisions were seen for conclusions to be drawn. Finally, advanced computing facilities were needed to process the vast amount of data (25 petabytes per year as of 2012) produced by the collisions. For the announcement of 4 July 2012, a new collider known as the Large Hadron Collider was constructed at CERN with a planned eventual collision energy of 14 TeV—over seven times any previous collider—and over 300 trillion (3×10^{14}) LHC proton–proton collisions were analysed by the LHC Computing Grid, the world's largest computing grid (as of 2012), comprising over 170 computing facilities in a worldwide network across 36 countries.

Search before 4 July 2012

The first extensive search for the Higgs boson was conducted at the Large Electron–Positron Collider (LEP) at CERN in the 1990s. At the end of its service in 2000, LEP had found no conclusive evidence for the Higgs. This implied that if the Higgs boson were to exist it would have to be heavier than 114.4 GeV/c^2.

The search continued at Fermilab in the United States, where the Tevatron—the collider that discovered the top quark in 1995—had been upgraded for this purpose. There was no guarantee that the Tevatron would be able to find the Higgs, but it was the only supercollider that was operational since the Large Hadron Collider (LHC) was still under construction and the planned Superconducting Super Collider had been cancelled in 1993 and never completed. The Tevatron was only able to exclude further ranges for the Higgs mass, and was shut down on 30 September 2011 because it no longer could keep up with the LHC. The final analysis of the data excluded the possibility of a Higgs boson with a mass between 147 GeV/c^2 and 180 GeV/c^2. In addition, there was a small (but not significant) excess of events possibly indicating a Higgs boson with a mass between 115 GeV/c^2 and 140 GeV/c^2.

The Large Hadron Collider at CERN in Switzerland, was designed specifically to be able to either confirm or exclude the existence of the Higgs boson. Built in a 27 km tunnel under the ground near Geneva originally inhabited by LEP, it was designed to collide two beams of protons, initially at energies of 3.5 TeV per beam (7 TeV total), or almost 3.6 times that of the Tevatron, and upgradeable to 2×7 TeV (14 TeV total) in future. Theory suggested if the Higgs boson existed, collisions at these energy levels should be able to reveal it. As one of the most complicated scientific instruments ever built, its operational readiness was delayed for 14 months by a magnet quench event nine days after its inaugural tests, caused by a faulty electrical connection that damaged over 50 superconducting magnets and contaminated the vacuum system.

Data collection at the LHC finally commenced in March 2010. By December 2011 the two main particle detectors at the LHC, ATLAS and CMS, had narrowed down the mass range where the Higgs could exist to around 116-130 GeV (ATLAS) and 115-127 GeV (CMS). There had also already been a number of promising event excesses that had «evaporated» and proven to be nothing but random fluctuations. However, from around May 2011, both experiments had seen among their results, the slow emergence of a small yet consistent excess of gamma and 4-lepton decay signatures and several other particle decays, all hinting at a new particle at a mass around 125 GeV. By around November 2011, the anomalous data at 125 GeV was becoming «too large to ignore» (although still far from conclusive), and the team leaders at both ATLAS and CMS each privately suspected they might have found the Higgs. On November 28, 2011, at an internal meeting of the two team leaders and the director general of CERN, the latest analyses were discussed outside their teams for the first time, suggesting both ATLAS and CMS might be converging on a possible shared result at 125 GeV, and initial preparations commenced in case of a successful finding. While this information was not known publicly at the time, the narrowing of the possible Higgs range to around 115–130 GeV and the repeated observation of small but consistent event excesses across multiple channels at both ATLAS and CMS in the 124-126 GeV region (described as «tantalising hints» of around 2-3 sigma) were public knowledge with «a lot of interest». It was therefore widely anticipated around the end of 2011, that the LHC would provide sufficient data to either exclude or confirm the finding of a Higgs boson by the end of 2012, when their 2012 collision data (with slightly higher 8 TeV collision energy) had been examined.

Discovery of Candidate Boson at CERN

On 22 June 2012 CERN announced an upcoming seminar covering tentative findings for 2012, and shortly afterwards (from around 1 July 2012 according to an analysis of the spreading rumour in social media) rumours began to spread in the media that this would include a major announcement, but it was unclear whether this would be a stronger signal or a formal discovery. Speculation escalated to a "fevered" pitch when reports emerged that Peter Higgs, who proposed the particle, was to be attending the seminar, and that "five leading physicists" had been invited – generally believed to signify the five living 1964 authors – with Higgs, Englert, Guralnik, Hagen attending and Kibble confirming his invitation (Brout having died in 2011).

On 4 July 2012 both of the CERN experiments announced they had independently made the same discovery: CMS of a previously unknown boson with mass 125.3 ± 0.6 GeV/c^2 and ATLAS of a boson with mass 126.0 ± 0.6 GeV/c^2. Using the combined analysis of two interaction types (known as 'channels'), both experiments independently reached a local significance of 5 sigma — implying that the probability of getting at least as strong a result by chance alone is less than 1 in 3 million. When additional channels were taken into account, the CMS significance was reduced to 4.9 sigma.

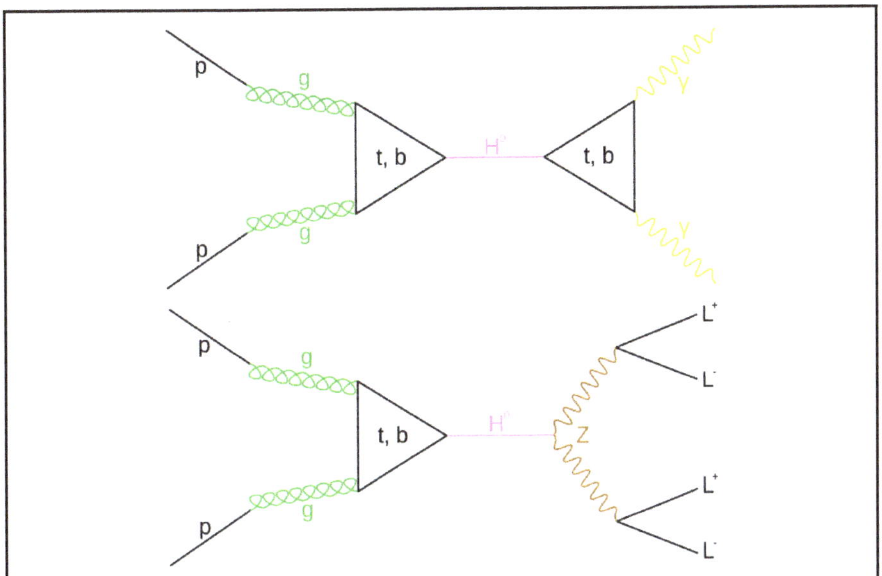

Feynman diagrams showing the cleanest channels associated with the low-mass (~125 GeV) Higgs boson candidate observed by ATLAS and CMS at the LHC. The dominant production mechanism at this mass involves two gluons from each proton fusing to a Top-quark Loop, which couples strongly to the Higgs field to produce a Higgs boson.

Left: Diphoton channel: Boson subsequently decays into 2 gamma ray photons by virtual interaction with a W boson loop or top quark loop.

Right: 4-Lepton "golden channel": Boson emits 2 Z bosons, which each decay into 2 leptons (electrons, muons).

Experimental analysis of these channels reached a significance of more than 5 sigma in both experiments.

The two teams had been working 'blinded' from each other from around late 2011 or early 2012, meaning they did not discuss their results with each other, providing additional certainty that any common finding was genuine validation of a particle. This level of evidence, confirmed independently by two separate teams and experiments, meets the formal level of proof required to announce a confirmed discovery.

On 31 July 2012, the ATLAS collaboration presented additional data analysis on the "observation of a new particle", including data from a third channel, which improved the significance to 5.9 sigma (1 in 588 million chance of obtaining at least as strong evidence by random background effects alone) and mass 126.0 ± 0.4 (stat) ± 0.4 (sys) GeV/c^2, and CMS improved the significance to 5-sigma and mass 125.3 ± 0.4 (stat) ± 0.5 (sys) GeV/c^2.

The New Particle Tested As A Possible Higgs boson

Following the 2012 Discovery, it was still unconfirmed whether or not the 125 GeV/c^2 particle was a Higgs boson. On one hand, observations remained consistent with the observed particle being the Standard Model Higgs boson, and the particle decayed into at least some of the predicted channels. Moreover, the production rates and branching ratios for the observed channels broadly matched the predictions by the Standard Model within the experimental uncertainties. However, the experimental uncertainties currently still left room for alternative explanations, meaning an announcement of the discovery of a Higgs boson would have been premature. To allow more opportunity for data collection, the LHC's proposed 2012 shutdown and 2013–14 upgrade were postponed by 7 weeks into 2013.

In November 2012, in a conference in Kyoto researchers said evidence gathered since July was falling into line with the basic Standard Model more than its alternatives, with a range of results for several interactions matching that theory's predictions. Physicist Matt Strassler highlighted "considerable" evidence that the new particle is not a pseudoscalar negative parity particle (consistent with this required finding for a Higgs boson), "evaporation" or lack of increased significance for previous hints of non-Standard Model findings, expected Standard Model interactions with W and Z bosons, absence of "significant new implications" for or against supersymmetry, and in general no significant deviations to date from the results expected of a Standard Model Higgs boson. However some kinds of extensions to the Standard Model would also show very similar results; so commentators noted that based on other particles that are still being understood long after their discovery, it may take years to be sure, and decades to fully understand the particle that has been found.

These findings meant that as of January 2013, scientists were very sure they had found an unknown particle of mass ~ 125 GeV/c^2, and had not been misled by experimental error or a chance result. They were also sure, from initial observations, that the new particle was some kind of boson. The behaviours and properties of the particle, so far as examined since July 2012, also seemed quite close to the behaviours expected of a Higgs boson. Even so, it could still have been a Higgs boson or some other unknown boson, since future tests could show behaviours that do not match a Higgs boson, so as of December 2012 CERN still only stated that the new particle was "consistent with" the Higgs boson, and scientists did not yet positively say it was the Higgs boson. Despite this, in late 2012, widespread media reports announced (incorrectly) that a Higgs boson had been confirmed during the year.

In January 2013, CERN director-general Rolf-Dieter Heuer stated that based on data analysis to date, an answer could be possible 'towards' mid-2013, and the deputy chair of physics at Brookhaven National Laboratory stated in February 2013 that a "definitive" answer might require "another few years" after the collider's 2015 restart. In early March 2013, CERN Research Director Sergio Bertolucci stated that confirming spin-0 was the major remaining requirement to determine whether the particle is at least some kind of Higgs boson.

Preliminary Confirmation of Existence and Current Status

On 14 March 2013 CERN confirmed that:

"CMS and ATLAS have compared a number of options for the spin-parity of this particle, and these all prefer no spin and even parity [two fundamental criteria of a Higgs boson consistent with the Standard Model]. This, coupled with the measured interactions of the new particle with other particles, strongly indicates that it is a Higgs boson."

This also makes the particle the first elementary scalar particle to be discovered in nature.

Examples of tests used to validate whether the 125 GeV particle is a Higgs boson:

Requirement	How tested / explanation	Current status (August 2016)
Zero spin	Examining decay patterns. Spin-1 had been ruled out at the time of initial discovery by the observed decay to two photons (γγ), leaving spin-0 and spin-2 as remaining candidates.	Spin-0 tentatively confirmed. The spin-2 hypothesis is excluded with a confidence level exceeding 99.9%.
Even (Positive) parity	Studying the angles at which decay products fly apart. Negative parity was also disfavoured if spin-0 was confirmed.	Even parity tentatively confirmed. The spin-0 negative parity hypothesis is excluded with a confidence level exceeding 99.9%.
Decay channels (outcomes of particle decaying) are as predicted	The Standard Model predicts the decay patterns of a 125–126 GeV Higgs boson. Are these all being seen, and at the right rates? Particularly significant, we should observe decays into pairs of photons (γγ), W and Z bosons (WW and ZZ), bottom quarks (bb), and tau leptons (ττ), among the possible outcomes.	γγ, ττ, WW and ZZ observed; bb not yet confirmed. All observed signal strengths are consistent with the Standard Model prediction.
Couples to mass (i.e., strength of interaction with Standard Model particles proportional to their mass)	Particle physicist Adam Falkowski states that the essential qualities of a Higgs boson are that it is a spin-0 (scalar) particle which also couples to mass (W and Z bosons); proving spin-0 alone is insufficient.	Couplings to mass strongly evidenced ("At 95% confidence level c_V is within 15% of the standard model value c_V=1").
Higher energy results remain consistent	After the LHC's 2015 restart at the LHC's full planned energies of 13–14 TeV, searches for multiple Higgs particles (as predicted in some theories) and tests targeting other versions of particle theory will take place. These higher energy results must continue to give results consistent with Higgs theories	Analysis of collisions up to July do not show deviations from the Standard Model, with experimental precisions approaching the earlier measurements at lower collision energy.

Public Discussion

Naming

Names Used by physicists

The name most strongly associated with the particle and field is the Higgs boson and Higgs field. For some time the particle was known by a combination of its PRL author names (including at times Anderson), for example the Brout–Englert–Higgs particle, the Anderson-Higgs particle, or the Englert–Brout–Higgs–Guralnik–Hagen–Kibble mechanism, and these are still used at times. Fuelled in part by the issue of recognition and a potential shared Nobel Prize, the most appropriate name was still occasionally a topic of debate until 2013. Higgs himself prefers to call the particle either by an acronym of all those involved, or "the scalar boson", or "the so-called Higgs particle".

A considerable amount has been written on how Higgs' name came to be exclusively used. Two main explanations are offered. The first is that Higgs undertook a step which was either unique, clearer or more explicit in his paper in formally predicting and examining the particle. Of the PRL papers' authors, only the paper by Higgs explicitly offered as a prediction that a massive particle would exist and calculated some of its properties; he was therefore "the first to postulate the existence of a massive particle" according to Nature. Physicist and author Frank Close and physicist-blogger Peter Woit both comment that the paper by GHK was also completed after Higgs and Brout–Englert were submitted to Physical Review Letters. and that Higgs alone had drawn attention to a predicted massive scalar boson, while all others had focused on the massive vector bosons; In this way, Higgs' contribution also provided experimentalists with a crucial "concrete target" needed to test the theory. However, in Higgs' view, Brout and Englert did not explicitly mention the boson since its existence is plainly obvious in their work, while according to Guralnik the GHK paper was a complete analysis of the entire symmetry breaking mechanism whose mathematical rigour is absent from the other two papers, and a massive particle may exist in some solutions. Higgs' paper also provided an "especially sharp" statement of the challenge and its solution according to science historian David Kaiser.

The alternative explanation is that the name was popularised in the 1970s due to its use as a convenient shorthand or because of a mistake in citing. Many accounts (including Higgs' own) credit the "Higgs" name to physicist Benjamin Lee (in Korean: Lee Whi-soh). Lee was a significant populist for the theory in its early stages, and habitually attached the name "Higgs" as a "convenient shorthand" for its components from 1972 and in at least one instance from as early as 1966. Although Lee clarified in his footnotes that "'Higgs' is an abbreviation for Higgs, Kibble, Guralnik, Hagen, Brout, Englert", his use of the term (and perhaps also Steven Weinberg's mistaken cite of Higgs' paper as the first in his seminal 1967 paper) meant that by around 1975–76 others had also begun to use the name 'Higgs' exclusively as a shorthand.

Nickname

The Higgs boson is often referred to as the "God particle" in popular media outside the scientific community. The nickname comes from the title of the 1993 book on the Higgs boson and particle physics, The God Particle: If the Universe Is the Answer, What Is the Question? by Nobel Physics prizewinner and Fermilab director Leon Lederman. Lederman wrote it in the context of failing US government support for the Superconducting Super Collider, a part-constructed titanic competitor to the Large Hadron Collider with planned collision energies of 2 × 20 TeV that was championed by Lederman since its 1983 inception and shut down in 1993. The book sought in part to promote awareness of the significance and need for such a project in the face of its possible loss of funding. Lederman, a leading researcher in the field, wanted to title his book The Goddamn Particle: If the Universe is the Answer, What is the Question? But his editor decided that the title was too controversial and convinced Lederman to change the title to The God Particle: If the Universe is the Answer, What is the Question?

While media use of this term may have contributed to wider awareness and interest, many scientists feel the name is inappropriate since it is sensational hyperbole and misleads readers; the particle also has nothing to do with God, leaves open numerous questions in fundamental physics, and does not explain the ultimate origin of the universe. Higgs, an atheist, was reported to be displeased and stated in a 2008 interview that he found it "embarrassing" because it was "the kind of misuse... which I think might offend some people". Science writer Ian Sample stated in his 2010 book on the search that the nickname is "universally hate[d]" by physicists and perhaps the "worst derided" in the history of physics, but that (according to Lederman) the publisher rejected all titles mentioning "Higgs" as unimaginative and too unknown.

Lederman begins with a review of the long human search for knowledge, and explains that his tongue-in-cheek title draws an analogy between the impact of the Higgs field on the fundamental symmetries at the Big Bang, and the apparent chaos of structures, particles, forces and interactions that resulted and shaped our present universe, with the biblical story of Babel in which the primordial single language of early Genesis was fragmented into many disparate languages and cultures.

Today ... we have the standard model, which reduces all of reality to a dozen or so particles and four forces. ... It's a hard-won simplicity [...and...] remarkably accurate. But it is also incomplete and, in fact, internally inconsistent... This boson is so central to the state of physics today, so crucial to our final understanding of the structure of matter, yet so elusive, that I have given it a nickname: the God Particle. Why God Particle? Two reasons. One, the publisher wouldn't let us call it the Goddamn Particle, though that might be a more appropriate title, given its villainous nature and the expense it is causing. And two, there is a connection, of sorts, to another book, a much older one...

— Leon M. Lederman and Dick Teresi, The God Particle: If the Universe is the Answer, What is the Question p. 22

Lederman asks whether the Higgs boson was added just to perplex and confound those seeking knowledge of the universe, and whether physicists will be confounded by it as recounted in that story, or ultimately surmount the challenge and understand "how beautiful is the universe [God has] made".

Other Proposals

A renaming competition by British newspaper The Guardian in 2009 resulted in their science correspondent choosing the name "the champagne bottle boson" as the best submission: "The bottom of a champagne bottle is in the shape of the Higgs potential and is often used as an illustration in physics lectures. So it's not an embarrassingly grandiose name, it is memorable, and [it] has some physics connection too." The name Higgson was suggested as well, in an opinion piece in the Institute of Physics' online publication physicsworld.com.

Media Explanations and Analogies

There has been considerable public discussion of analogies and explanations for the Higgs particle and how the field creates mass, including coverage of explanatory attempts in their own right and a competition in 1993 for the best popular explanation by then-UK Minister for Science Sir William Waldegrave and articles in newspapers worldwide.

Photograph of light passing through a dispersive prism: the rainbow effect arises because photons are not all affected to the same degree by the dispersive material of the prism.

An educational collaboration involving an LHC physicist and a High School Teachers at CERN educator suggests that dispersion of light – responsible for the rainbow and dispersive prism – is a useful analogy for the Higgs field's symmetry breaking and mass-causing effect.

Symmetry breaking in optics	In a vacuum, light of all colours (or photons of all wavelengths) travels at the same velocity, a symmetrical situation. In some substances such as glass, water or air, this symmetry is broken (See: Photons in matter). The result is that light of different wavelengths appears to have different velocities (as seen from outside).
Symmetry breaking in particle physics	In 'naive' gauge theories, gauge bosons and other fundamental particles are all massless – also a symmetrical situation. In the presence of the Higgs field this symmetry is broken. The result is that particles of different types will have different masses.

Matt Strassler uses electric fields as an analogy:

Some particles interact with the Higgs field while others don't. Those particles that feel the Higgs field act as if they have mass. Something similar happens in an electric field – charged objects are pulled around and neutral objects can sail through unaffected. So you can think of the Higgs search as an attempt to make waves in the Higgs field [create Higgs bosons] to prove it's really there.

A similar explanation was offered by The Guardian:

The Higgs boson is essentially a ripple in a field said to have emerged at the birth of the universe and to span the cosmos to this day ... The particle is crucial however: it is the smoking gun, the evidence required to show the theory is right.

The Higgs field's effect on particles was famously described by physicist David Miller as akin to a room full of political party workers spread evenly throughout a room: the crowd gravitates to and slows down famous people but does not slow down others. He also drew attention to well-known effects in solid state physics where an electron's effective mass can be much greater than usual in the presence of a crystal lattice.

Analogies based on drag effects, including analogies of "syrup" or "molasses" are also well known, but can be somewhat misleading since they may be understood (incorrectly) as saying that the Higgs field simply resists some particles' motion but not others' – a simple resistive effect could also conflict with Newton's third law.

Recognition and Awards

There has been considerable discussion of how to allocate the credit if the Higgs boson is proven, made more pointed as a Nobel prize had been expected, and the very wide basis of people entitled to consideration. These include a range of theoreticians who made the Higgs mechanism theory possible, the theoreticians of the 1964 PRL papers (including Higgs himself), the theoreticians who derived from these, a working electroweak theory and the Standard Model itself, and also the experimentalists at CERN and other institutions who made possible the proof of the Higgs field and boson in reality. The Nobel prize has a limit of 3 persons to share an award, and some possible winners are already prize holders for other work, or are deceased (the prize is only awarded to persons in their lifetime). Existing prizes for works relating to the Higgs field, boson, or mechanism include:

- Nobel Prize in Physics (1979) – Glashow, Salam, and Weinberg, for contributions to the theory of the unified weak and electromagnetic interaction between elementary particles
- Nobel Prize in Physics (1999) – 't Hooft and Veltman, for elucidating the quantum structure of electroweak interactions in physics
- Nobel Prize in Physics (2008) – Nambu (shared), for the discovery of the mechanism of spontaneous broken symmetry in subatomic physics
- J. J. Sakurai Prize for Theoretical Particle Physics (2010) – Hagen, Englert, Guralnik, Higgs, Brout, and Kibble, for elucidation of the properties of spontaneous symmetry breaking in four-dimensional relativistic gauge theory and of the mechanism for the consistent generation of vector boson masses (for the 1964 papers described above)
- Wolf Prize (2004) – Englert, Brout, and Higgs
- Nobel Prize in Physics (2013) – Peter Higgs and François Englert, for the theoretical discovery of a mechanism that contributes to our understanding of the origin of mass of subatomic particles, and which recently was confirmed through the discovery of the predicted fundamental particle, by the ATLAS and CMS experiments at CERN's Large Hadron Collider

Additionally Physical Review Letters' 50-year review (2008) recognised the 1964 PRL symmetry breaking papers and Weinberg's 1967 paper A model of Leptons (the most cited paper in particle physics, as of 2012) "milestone Letters".

Following reported observation of the Higgs-like particle in July 2012, several Indian media outlets reported on the supposed neglect of credit to Indian physicist Satyendra Nath Bose after whose work in the 1920s the class of particles "bosons" is named (although physicists have described Bose's connection to the discovery as tenuous).

REFERENCES

1. Wells, John C. (1990). Longman pronunciation dictionary. Harlow, England: Longman. ISBN 0582053838. entry "Boson"
2. "boson". Collins Dictionary.
3. Carroll, Sean (2007) Dark Matter, Dark Energy: The Dark Side of the Universe, Guidebook Part 2 p. 43, The Teaching Company, ISBN 1598033506 "...boson: A force-carrying particle, as opposed to a matter particle (fermion). Bosons can be piled on top of each other without limit. Examples include photons, gluons, gravitons, weak bosons, and the Higgs boson. The spin of a boson is always an integer, such as 0, 1, 2, and so on..."
4. Notes on Dirac's lecture Developments in Atomic Theory at Le Palais de la Découverte, 6 December 1945, UKNATARCHI Dirac Papers BW83/2/257889. See note 64 to p. 331 in "The Strangest Man" by Graham Farmelo
5. Daigle, Katy (10 July 2012). "India: Enough about Higgs, let's discuss the boson". AP News. Retrieved 10 July 2012.

6. Bal, Hartosh Singh (19 September 2012). "The Bose in the Boson". New York Times blog. Retrieved 21 September 2012.
7. "Higgs boson: The poetry of subatomic particles". BBC News. 4 July 2012. Retrieved 6 July 2012.
8. Charles P. Poole, Jr. (11 March 2004). Encyclopedic Dictionary of Condensed Matter Physics. Academic Press. ISBN 978-0-08-054523-3.
9. "boson". Merriam-Webster Online Dictionary. Retrieved 21 March 2010.
10. Carroll, Sean. "Explain it in 60 seconds: Bosons". Symmetry Magazine. Fermilab/ SLAC. Retrieved 15 February 2013.
11. Srednicki, Mark (2007). Quantum Field Theory, Cambridge University Press, pp. 28–29, ISBN 978-0-521-86449-7.
12. Sakurai, J.J. (1994). Modern Quantum Mechanics (Revised Edition), p. 362. Addison-Wesley, ISBN 0-201-53929-2.
13. Gribbin, John (2000). Q is for Quantum – An Encyclopedia of Particle Physics. Simon & Schuster. ISBN 0-684-85578-X.
14. Clark, John, E.O. (2004). The Essential Dictionary of Science. Barnes & Noble. ISBN 0-7607-4616-8.
15. Veltman, Martinus (2003). Facts and Mysteries in Elementary Particle Physics. World Scientific. ISBN 981-238-149-X.
16. "CERN and the Higgs boson". CERN. Retrieved 23 November 2016.
17. O'Luanaigh, C. (14 March 2013). "New results indicate that new particle is a Higgs boson". CERN. Retrieved 2013-10-09.
18. Bryner, J. (14 March 2013). "Particle confirmed as Higgs boson". NBC News. Retrieved 2013-03-14.
19. Heilprin, J. (14 March 2013). "Higgs Boson Discovery Confirmed After Physicists Review Large Hadron Collider Data at CERN". The Huffington Post. Retrieved 2013-03-14.
20. ATLAS; CMS (26 March 2015). "Combined Measurement of the Higgs Boson Mass in pp Collisions at $\sqrt{s}$=7 and 8 TeV with the ATLAS and CMS Experiments". arXiv:1503.07589Freely accessible.
21. LHC Higgs Cross Section Working Group; Dittmaier; Mariotti; Passarino; Tanaka; Alekhin; Alwall; Bagnaschi; Banfi (2012). "Handbook of LHC Higgs Cross Sections: 2. Differential Distributions". CERN Report 2 (Tables A.1 – A.20). 1201: 3084. arXiv:1201.3084Freely accessible. Bibcode:2012arXiv1201.3084L.
22. Onyisi, P. (23 October 2012). "Higgs boson FAQ". University of Texas ATLAS group. Retrieved 2013-01-08.
23. Strassler, M. (12 October 2012). "The Higgs FAQ 2.0". ProfMattStrassler.com. Retrieved 2013-01-08. [Q] Why do particle physicists care so much about the Higgs particle?
24. [A] Well, actually, they don't. What they really care about is the Higgs field, because it is so important. [emphasis in original]

25. José Luis Lucio; Arnulfo Zepeda (1987). Proceedings of the II Mexican School of Particles and Fields, Cuernavaca-Morelos, 1986. World Scientific. p. 29. ISBN 9971504340.
26. Gunion, Dawson, Kane, and Haber (199). The Higgs Hunter's Guide (1st ed.). pp. 11 (?). ISBN 978-0-7867-4318-6. – quoted as being in the first (1990) edition of the book by Peter Higgs in his talk "My Life as a Boson", 2001, ref#25.
27. Strassler, M. (8 October 2011). "The Known Particles – If The Higgs Field Were Zero". ProfMattStrassler.com. Retrieved 13 November 2012. The Higgs field: so important it merited an entire experimental facility, the Large Hadron Collider, dedicated to understanding it.
28. Biever, C. (6 July 2012). "It's a boson! But we need to know if it's the Higgs". New Scientist. Retrieved 2013-01-09.
29. Q: 'If we don't know the new particle is a Higgs, what do we know about it?' We know it is some kind of boson, says Vivek Sharma of CMS [...]
30. Q: 'are the CERN scientists just being too cautious? What would be enough evidence to call it a Higgs boson?' As there could be many different kinds of Higgs bosons, there's no straight answer.
31. Siegfried, T. (20 July 2012). "Higgs Hysteria". Science News. Retrieved 2012-12-09. In terms usually reserved for athletic achievements, news reports described the finding as a monumental milestone in the history of science.
32. Del Rosso, A. (19 November 2012). "Higgs: The beginning of the exploration". CERN Bulletin. Retrieved 2013-01-09. Even in the most specialized circles, the new particle discovered in July is not yet being called the "Higgs boson". Physicists still hesitate to call it that before they have determined that its properties fit with those the Higgs theory predicts the Higgs boson has.
33. Naik, G. (14 March 2013). "New Data Boosts Case for Higgs Boson Find". The Wall Street Journal. Retrieved 2013-03-15. 'We've never seen an elementary particle with spin zero,' said Tony Weidberg, a particle physicist at the University of Oxford who is also involved in the CERN experiments.
34. Overbye, D. (8 October 2013). "For Nobel, They Can Thank the 'God Particle'". The New York Times. Retrieved 2013-11-03.
35. Sample, I. (29 May 2009). "Anything but the God particle". The Guardian. Retrieved 2009-06-24.
36. Evans, R. (14 December 2011). "The Higgs boson: Why scientists hate that you call it the 'God particle'". National Post. Retrieved 2013-11-03.
37. The nickname occasionally has been satirised in mainstream media as well. Borowitz, Andy (July 13, 2012). "5 questions for the Higgs boson". The New Yorker.
38. Leon M. Lederman; Dick Teresi (1993). The God Particle: If the Universe is the Answer, What is the Question. Houghton Mifflin Company.
39. Heath, Nick, The Cern tech that helped track down the God particle, TechRepublic, 4 July 2012

40. Griffiths, David (2008), Introduction to Elementary Particles (2nd revised ed.), WILEY-VCH, ISBN 978-3-527-40601-2

41. Shu, F. H. (1982). The Physical Universe: An Introduction to Astronomy. University Science Books. pp. 107–108. ISBN 978-0-935702-05-7.

42. Rao, Achintya (2 July 2012). "Why would I care about the Higgs boson?". CMS Public Website. CERN. Retrieved 18 July 2012.

43. Max Jammer, Concepts of Mass in Contemporary Physics and Philosophy (Princeton, NJ: Princeton University Press, 2000) pp.162–163, who provides many references in support of this statement.

44. Alekhin, Djouadi and Moch, S.; Djouadi, A.; Moch, S. (2012-08-13). "The top quark and Higgs boson masses and the stability of the electroweak vacuum". Physics Letters B. 716: 214–219. arXiv:1207.0980Freely accessible. Bibcode:2012PhLB..716..214A. doi:10.1016/j.physletb.2012.08.024. Retrieved 20 February 2013.

45. M.S. Turner; F. Wilczek (1982). "Is our vacuum metastable?". Nature. 298 (5875): 633–634. Bibcode:1982Natur.298..633T. doi:10.1038/298633a0.

46. S. Coleman; F. De Luccia (1980). "Gravitational effects on and of vacuum decay". Physical Review. D21 (12): 3305–3315. Bibcode:1980PhRvD..21.3305C. doi:10.1103/PhysRevD.21.3305.

47. M. Stone (1976). "Lifetime and decay of excited vacuum states". Phys. Rev. D. 14 (12): 3568–3573. Bibcode:1976PhRvD..14.3568S. doi:10.1103/PhysRevD.14.3568.

48. P.H. Frampton (1976). "Vacuum Instability and Higgs Scalar Mass". Phys. Rev. Lett. 37 (21): 1378–1380. Bibcode:1976PhRvL..37.1378F. doi:10.1103/PhysRevLett.37.1378.

49. P.H. Frampton (1977). "Consequences of Vacuum Instability in Quantum Field Theory". Phys. Rev. D15 (10): 2922–28. Bibcode:1977PhRvD..15.2922F. doi:10.1103/PhysRevD.15.2922.

50. Ellis, Espinosa, Giudice, Hoecker, & Riotto, J.; Espinosa, J.R.; Giudice, G.F.; Hoecker, A.; Riotto, A. (2009). "The Probable Fate of the Standard Model". Phys. Lett. B. 679 (4): 369–375. arXiv:0906.0954Freely accessible. Bibcode:2009PhLB..679..369E. doi:10.1016/j.physletb.2009.07.054.

51. Masina, Isabella (2013-02-12). "Higgs boson and top quark masses as tests of electroweak vacuum stability". Phys. Rev. D. 87 (5): 53001. arXiv:1209.0393Freely accessible. Bibcode:2013PhRvD..87e3001M. doi:10.1103/PhysRevD.87.053001.

52. Buttazzo, Degrassi, Giardino, Giudice, Sala, Salvio, Strumia (2013). "Investigating the near-criticality of the Higgs boson". JHEP. 2013: 089. arXiv:1307.3536Freely accessible. Bibcode:2013JHEP...12..089B. doi:10.1007/JHEP12(2013)089.

53. Irene Klotz (editing by David Adams and Todd Eastham) (2013-02-18). "Universe Has Finite Lifespan, Higgs Boson Calculations Suggest". Huffington Post. Reuters.

54. Hoffman, Mark (2013-02-19). "Higgs Boson Will Destroy The Universe Eventually". ScienceWorldReport. Retrieved 21 February 2013.

55. "Higgs boson will aid in creation of the universe – and how it will end". Catholic Online/NEWS CONSORTIUM. 2013-02-20. Retrieved 21 February 2013. [T]he Earth will likely be long gone before any Higgs boson particles set off an apocalyptic assault on the universe
56. Salvio, Alberto (2015-04-09). "A Simple Motivated Completion of the Standard Model below the Planck Scale: Axions and Right-Handed Neutrinos". Physics Letters B. 743: 428–434. arXiv:1501.03781Freely accessible. Bibcode:2015PhLB..743..428S. doi:10.1016/j.physletb.2015.03.015.
57. Boyle, Alan (2013-02-19). "Will our universe end in a 'big slurp'? Higgs-like particle suggests it might". NBC News' Cosmic log. Retrieved 21 February 2013.
58. Peralta, Eyder (2013-02-19). "If Higgs Boson Calculations Are Right, A Catastrophic 'Bubble' Could End Universe". npr – two way. Retrieved 21 February 2013.
59. Bezrukov; Shaposhnikov (2008-01-24). "The Standard Model Higgs boson as the inflaton". Phys. Lett. B. 659: 703–706. arXiv:0710.3755Freely accessible. Bibcode:2008PhLB..659..703B. doi:10.1016/j.physletb.2007.11.072.
60. Salvio, Alberto (2013-08-09). "Higgs Inflation at NNLO after the Boson Discovery". Phys. Lett. B. 727: 234–239. arXiv:1308.2244Freely accessible. Bibcode:2013PhLB..727..234S. doi:10.1016/j.physletb.2013.10.042.
61. Cole, K. (2000-12-14). "One Thing Is Perfectly Clear: Nothingness Is Perfect". Los Angeles Times. p. 'Science File'. Retrieved 17 January 2013.
62. Why the public should care about the Higgs Boson – Jodi Lieberman, American Physical Society (APS)
63. Matt Strassler's blog – Why the Higgs particle matters 2 July 2012
64. Sean Carroll (13 November 2012). The Particle at the End of the Universe: How the Hunt for the Higgs Boson Leads Us to the Edge of a New World. Penguin Group US. ISBN 978-1-101-60970-5.
65. Woit, Peter (13 November 2010). "The Anderson–Higgs Mechanism". Dr. Peter Woit (Senior Lecturer in Mathematics Columbia University and Ph.D. particle physics). Retrieved 12 November 2012.
66. Goldstone, J; Salam, Abdus; Weinberg, Steven (1962). "Broken Symmetries". Physical Review. 127 (3): 965–970. Bibcode:1962PhRv..127..965G. doi:10.1103/PhysRev.127.965.
67. Guralnik, G. S. (2011). "The Beginnings of Spontaneous Symmetry Breaking in Particle Physics". arXiv:1110.2253Freely accessible [physics.hist-ph].
68. Kibble, T. W. B. (2009). "Englert–Brout–Higgs–Guralnik–Hagen–Kibble Mechanism". Scholarpedia. 4 (1): 6441. Bibcode:2009SchpJ...4.6441K. doi:10.4249/scholarpedia.6441. Retrieved 2012-11-23.
69. Kibble, T. W. B. "History of Englert–Brout–Higgs–Guralnik–Hagen–Kibble Mechanism (history)". Scholarpedia. 4 (1): 8741. Bibcode:2009SchpJ...4.8741K. doi:10.4249/scholarpedia.8741. Retrieved 2012-11-23.
70. The Nobel Prize in Physics 2008 – official Nobel Prize website.

71. List of Anderson 1958–1959 papers referencing 'symmetry', at APS Journals
72. Higgs, Peter (2010-11-24). "My Life as a Boson" (PDF). Talk given by Peter Higgs at Kings College, London, Nov 24 2010, expanding on a paper originally presented in 2001.
73. Anderson, P. (1963). "Plasmons, gauge invariance and mass". Physical Review. 130: 439–442. Bibcode:1963PhRv..130..439A. doi:10.1103/PhysRev.130.439.
74. Klein, A.; Lee, B. (1964). "Does Spontaneous Breakdown of Symmetry Imply Zero-Mass Particles?". Physical Review Letters. 12 (10): 266–268. Bibcode:1964PhRvL..12..266K. doi:10.1103/PhysRevLett.12.266.
75. Englert, François; Brout, Robert (1964). "Broken Symmetry and the Mass of Gauge Vector Mesons". Physical Review Letters. 13 (9): 321–23. Bibcode:1964PhRvL..13..321E. doi:10.1103/PhysRevLett.13.321.
76. Higgs, Peter (1964). "Broken Symmetries and the Masses of Gauge Bosons". Physical Review Letters. 13 (16): 508–509. Bibcode:1964PhRvL..13..508H. doi:10.1103/PhysRevLett.13.508.
77. Guralnik, Gerald; Hagen, C. R.; Kibble, T. W. B. (1964). "Global Conservation Laws and Massless Particles". Physical Review Letters. 13 (20): 585–587. Bibcode:1964PhRvL..13..585G. doi:10.1103/PhysRevLett.13.585.
78. Higgs, Peter (1964). "Broken symmetries, massless particles and gauge fields". Physics Letters. 12 (2): 132–133. Bibcode:1964PhL....12..132H. doi:10.1016/0031-9163(64)91136-9.
79. Higgs, Peter (2010-11-24). "My Life as a Boson" (PDF). Talk given by Peter Higgs at Kings College, London, Nov 24 2010.
80. G.S. Guralnik (2011). "Gauge invariance and the Goldstone theorem – 1965 Feldafing talk". Modern Physics Letters A. 26 (19): 1381–1392. arXiv:1107.4592Freely accessible. Bibcode:2011MPLA...26.1381G. doi:10.1142/S0217732311036188.
81. Higgs, Peter (1966). "Spontaneous Symmetry Breakdown without Massless Bosons". Physical Review. 145 (4): 1156–1163. Bibcode:1966PhRv..145.1156H. doi:10.1103/PhysRev.145.1156.
82. Kibble, Tom (1967). "Symmetry Breaking in Non-Abelian Gauge Theories". Physical Review. 155 (5): 1554–1561. Bibcode:1967PhRv..155.1554K. doi:10.1103/PhysRev.155.1554.
83. "Guralnik, G S; Hagen, C R and Kibble, T W B (1967). Broken Symmetries and the Goldstone Theorem. Advances in Physics, vol. 2" (PDF).
84. "Physical Review Letters – 50th Anniversary Milestone Papers". Physical Review Letters.
85. S. Weinberg (1967). "A Model of Leptons". Physical Review Letters. 19 (21): 1264–1266. Bibcode:1967PhRvL..19.1264W. doi:10.1103/PhysRevLett.19.1264.
86. A.Salam (1968). N. Svartholm, ed. Elementary Particle Physics: Relativistic Groups and Analyticity. Eighth Nobel Symposium. Stockholm: Almquvist and Wiksell. p. 367.

87. S.L. Glashow (1961). "Partial-symmetries of weak interactions". Nuclear Physics. 22 (4): 579–588. Bibcode:1961NucPh..22..579G. doi:10.1016/0029-5582(61)90469-2.
88. Ellis, John; Gaillard, Mary K.; Nanopoulos, Dimitri V. (2012). "A Historical Profile of the Higgs Boson". arXiv:1201.6045Freely accessible [hep-ph].
89. "Martin Veltman Nobel Lecture, December 12, 1999, p.391" (PDF). Retrieved 2013-10-09.
90. Politzer, David. "The Dilemma of Attribution". Nobel Prize lecture, 2004.
91. Coleman, Sidney (1979-12-14). "The 1979 Nobel Prize in Physics". Science. 206 (4424): 1290–1292. Bibcode:1979Sci...206.1290C. doi:10.1126/science.206.4424.1290.
92. Letters from the Past – A PRL Retrospective (50 year celebration, 2008)
93. Jeremy Bernstein (January 1974). "Spontaneous symmetry breaking, gauge theories, the Higgs mechanism and all that" (PDF). Reviews of Modern Physics. 46 (1): 7–48. Bibcode:1974RvMP...46....7B. doi:10.1103/RevModPhys.46.7. Retrieved 2012-12-10.
94. American Physical Society – "J. J. Sakurai Prize for Theoretical Particle Physics".
95. Merali, Zeeya (4 August 2010). "Physicists get political over Higgs". Nature Magazine. Retrieved 28 December 2011.
96. Close, Frank (2011). The Infinity Puzzle: Quantum Field Theory and the Hunt for an Orderly Universe. Oxford: Oxford University Press. ISBN 978-0-19-959350-7.
97. G.S. Guralnik (2009). "The History of the Guralnik, Hagen and Kibble development of the Theory of Spontaneous Symmetry Breaking and Gauge Particles". International Journal of Modern Physics A. 24 (14): 2601–2627. arXiv:0907.3466Freely accessible. Bibcode:2009IJMPA..24.2601G. doi:10.1142/S0217751X09045431.
98. Peskin, Michael E.; Schroeder, Daniel V. (1995). Introduction to Quantum Field Theory. Reading, MA: Addison-Wesley Publishing Company. pp. 717–719 and 787–791. ISBN 0-201-50397-2.
99. Gunion, John (2000). The Higgs Hunter's Guide (illustrated, reprint ed.). Westview Press. pp. 1–3. ISBN 978-0-7382-0305-8.
100. Lisa Randall, Warped Passages: Unraveling the Mysteries of the Universe's Hidden Dimensions, p. 286: "People initially thought of tachyons as particles travelling faster than the speed of light ... But we now know that a tachyon indicates an instability in a theory that contains it. Regrettably for science fiction fans, tachyons are not real physical particles that appear in nature."
101. Sen, Ashoke (April 2002). "Rolling Tachyon". J. High Energy Phys. 2002 (0204): 048. arXiv:hep-th/0203211Freely accessible. Bibcode:2002JHEP...04..048S. doi:10.1088/1126-6708/2002/04/048.
102. Kutasov, David; Marino, Marcos & Moore, Gregory W. (2000). "Some exact results on tachyon condensation in string field theory". JHEP. 0010: 045. arXiv:hep-th/0009148Freely accessible. Bibcode:2000JHEP...10..045K. doi:10.1088/1126-

6708/2000/10/045.

103. Aharonov, Y.; Komar, A.; Susskind, L. (1969). "Superluminal Behavior, Causality, and Instability". Phys. Rev. American Physical Society. 182 (5): 1400–1403. Bibcode:1969PhRv..182.1400A. doi:10.1103/PhysRev.182.1400.
104. Feinberg, Gerald (1967). "Possibility of Faster-Than-Light Particles". Physical Review. 159 (5): 1089–1105. Bibcode:1967PhRv..159.1089F. doi:10.1103/PhysRev.159.1089.
105. Michael E. Peskin and Daniel V. Schroeder (1995). An Introduction to Quantum Field Theory, Perseus books publishing.
106. Flatow, Ira (6 July 2012). "At Long Last, The Higgs Particle... Maybe". NPR. Retrieved 10 July 2012.
107. "Explanatory Figures for the Higgs Boson Exclusion Plots". ATLAS News. CERN. Retrieved 6 July 2012.
108. Bernardi, G.; Carena, M.; Junk, T. (2012). "Higgs Bosons: Theory and Searches" (PDF). p. 7.
109. Lykken, Joseph D. (2009). "Beyond the Standard Model". Proceedings of the 2009 European School of High-Energy Physics, Bautzen, Germany, 14 – 27 June 2009. arXiv:1005.1676Freely accessible.
110. Plehn, Tilman (2012). Lectures on LHC Physics. Lecture Notes in Physics. 844. Springer. Sec. 1.2.2. arXiv:0910.4182Freely accessible. ISBN 3-642-24039-9.
111. "LEP Electroweak Working Group".
112. Peskin, Michael E.; Wells, James D. (2001). "How Can a Heavy Higgs Boson be Consistent with the Precision Electroweak Measurements?". Physical Review D. 64 (9): 093003. arXiv:hep-ph/0101342Freely accessible. Bibcode:2001PhRvD..64i3003P. doi:10.1103/PhysRevD.64.093003.
113. Baglio, Julien; Djouadi, Abdelhak (2011). "Higgs production at the lHC". Journal of High Energy Physics. 1103 (3): 055. arXiv:1012.0530Freely accessible. Bibcode:2011JHEP...03..055B. doi:10.1007/JHEP03(2011)055.
114. Baglio, Julien; Djouadi, Abdelhak (2010). "Predictions for Higgs production at the Tevatron and the associated uncertainties". Journal of High Energy Physics. 1010 (10): 063. arXiv:1003.4266Freely accessible. Bibcode:2010JHEP...10..064B. doi:10.1007/JHEP10(2010)064.
115. Teixeira-Dias (LEP Higgs working group), P. (2008). "Higgs boson searches at LEP". Journal of Physics: Conference Series. 110 (4): 042030. arXiv:0804.4146Freely accessible. Bibcode:2008JPhCS.110d2030T. doi:10.1088/1742-6596/110/4/042030.
116. Asquith, Lily (22 June 2012). "Why does the Higgs decay?". Life and Physics. London: The Guardian. Retrieved 14 August 2012.
117. "Higgs bosons: theory and searches" (PDF). PDGLive. Particle Data Group. 12 July 2012. Retrieved 15 August 2012.
118. Branco, G. C.; Ferreira, P.M.; Lavoura, L.; Rebelo, M.N.; Sher, Marc; Silva, João P. (July 2012). "Theory and phenomenology of two-Higgs-doublet models". Physics Reports. Elsevier. 516 (1): 1–102. arXiv:1106.0034Freely accessible.

Bibcode:2012PhR...516....1B. doi:10.1016/j.physrep.2012.02.002.

119. Csaki, C.; Grojean, C.; Pilo, L.; Terning, J. (2004). "Towards a realistic model of Higgsless electroweak symmetry breaking". Physical Review Letters. 92 (10): 101802. arXiv:hep-ph/0308038Freely accessible. Bibcode:2004PhRvL..92j1802C. doi:10.1103/PhysRevLett.92.101802. PMID 15089195.

120. Csaki, C.; Grojean, C.; Pilo, L.; Terning, J.; Terning, John (2004). "Gauge theories on an interval: Unitarity without a Higgs". Physical Review D. 69 (5): 055006. arXiv:hep-ph/0305237Freely accessible. Bibcode:2004PhRvD..69e5006C. doi:10.1103/PhysRevD.69.055006.

121. "The Hierarchy Problem: why the Higgs has a snowball's chance in hell". Quantum Diaries. 2012-07-01. Retrieved 19 March 2013.

122. "The Hierarchy Problem | Of Particular Significance". Profmattstrassler.com. Retrieved 2013-10-09.

123. "Collisions". LHC Machine Outreach. CERN. Retrieved 26 July 2012.

124. "Hunt for Higgs boson hits key decision point". MSNBC. 2012-12-06. Retrieved 2013-01-19.

125. Worldwide LHC Computing Grid main page 14 November 2012: "[A] global collaboration of more than 170 computing centres in 36 countries ... to store, distribute and analyse the ~25 Petabytes (25 million Gigabytes) of data annually generated by the Large Hadron Collider"

126. What is the Worldwide LHC Computing Grid? (Public 'About' page) 14 November 2012: "Currently WLCG is made up of more than 170 computing centers in 36 countries...The WLCG is now the world's largest computing grid"

127. W.-M. Yao; et al. (2006). "Review of Particle Physics" (PDF). Journal of Physics G. 33: 1–1232. arXiv:astro-ph/0601168Freely accessible. Bibcode:2006JPhG...33....1Y. doi:10.1088/0954-3899/33/1/001.

128. The CDF Collaboration, the D0 Collaboration, the Tevatron New Physics, Higgs Working Group (2012). "Updated Combination of CDF and D0 Searches for Standard Model Higgs Boson Production with up to 10.0 fb−1 of Data". arXiv:1207.0449Freely accessible [hep-ex].

129. "Interim Summary Report on the Analysis of the 19 September 2008 Incident at the LHC" (PDF). CERN. 15 October 2008. EDMS 973073. Retrieved 28 September 2009.

130. "CERN releases analysis of LHC incident" (Press release). CERN Press Office. 16 October 2008. Retrieved 12 November 2016.

131. "LHC to restart in 2009" (Press release). CERN Press Office. 5 December 2008. Retrieved 12 November 2016.

132. "LHC progress report". The Bulletin. CERN. 3 May 2010. Retrieved 7 December 2011.

133. "ATLAS experiment presents latest Higgs search status". ATLAS homepage. CERN. 13 December 2011. Retrieved 13 December 2011.

134. Taylor, Lucas (13 December 2011). "CMS search for the Standard Model

Higgs Boson in LHC data from 2010 and 2011". CMS public website. CERN. Retrieved 13 December 2011.

135. Overbye, D. (5 March 2013). "Chasing The Higgs Boson". The New York Times. Retrieved 2013-03-05.
136. "ATLAS and CMS experiments present Higgs search status" (Press release). CERN Press Office. 13 December 2011.
137. CMS collaboration (2014). "Precise determination of the mass of the Higgs boson and tests of compatibility of its couplings with the standard model predictions using proton collisions at 7 and 8 TeV". arXiv:1412.8662Freely accessible.
138. ATLAS collaboration (2014). "Measurements of Higgs boson production and couplings in the four-lepton channel in pp collisions at center-of-mass energies of 7 and 8 TeV with the ATLAS detector". arXiv:1408.5191Freely accessible.
139. ATLAS collaboration (2014). "Measurement of Higgs boson production in the diphoton decay channel in pp collisions at center-of-mass energies of 7 and 8 TeV with the ATLAS detector". arXiv:1408.7084Freely accessible.
140. "Press Conference: Update on the search for the Higgs boson at CERN on 4 July 2012". Indico.cern.ch. 22 June 2012. Retrieved 4 July 2012.
141. "CERN to give update on Higgs search as curtain raiser to ICHEP conference". CERN. 22 June 2012. Retrieved 12 November 2016.
142. "Scientists analyse global Twitter gossip around Higgs boson discovery". phys. org (from arXiv). 2013-01-23.
143. "Higgs boson particle results could be a quantum leap". Times LIVE. 28 June 2012. Retrieved 4 July 2012.
144. CERN prepares to deliver Higgs particle findings, Australian Broadcasting Corporation. Retrieved 4 July 2012.
145. "God Particle Finally Discovered? Higgs Boson News At Cern Will Even Feature Scientist It's Named After". Huffingtonpost.co.uk. Retrieved 2013-01-19.
146. Our Bureau (2012-07-04). "Higgs on way, theories thicken". Calcutta, India: Telegraphindia.com. Retrieved 2013-01-19.
147. Thornhill, Ted (2013-07-03). "God Particle Finally Discovered? Higgs Boson News At Cern Will Even Feature Scientist It's Named After". Huffington Post. Retrieved 23 July 2013.
148. Cooper, Rob (2013-07-01) [updated subsequently]. "God particle is 'found': Scientists at Cern expected to announce on Wednesday Higgs boson particle has been discovered". Daily Mail. London. Retrieved 23 July 2013. - States that ""Five leading theoretical physicists have been invited to the event on Wednesday – sparking speculation that the particle has been discovered.", including Higgs and Englert, and that Kibble – who was invited but unable to attend - "told the Sunday Times: 'My guess is that is must be a pretty positive result for them to be asking us out there'."
149. Adrian Cho (13 July 2012). "Higgs Boson Makes Its Debut After Decades-Long Search". Science. 337 (6091): 141–143. doi:10.1126/science.337.6091.141.

PMID 22798574.

150. CMS collaboration (2012). "Observation of a new boson at a mass of 125 GeV with the CMS experiment at the LHC". Physics Letters B. 716 (1): 30–61. arXiv:1207.7235Freely accessible. Bibcode:2012PhLB..716...30C. doi:10.1016/j.physletb.2012.08.021.
151. Taylor, Lucas (4 July 2012). "Observation of a New Particle with a Mass of 125 GeV". CMS Public Website. CERN. Retrieved 4 July 2012.
152. "Latest Results from ATLAS Higgs Search". ATLAS News. CERN. 4 July 2012. Retrieved 4 July 2012.
153. ATLAS collaboration (2012). "Observation of a New Particle in the Search for the Standard Model Higgs Boson with the ATLAS Detector at the LHC". Physics Letters B. 716 (1): 1–29. arXiv:1207.7214Freely accessible. Bibcode:2012PhLB..716....1A. doi:10.1016/j.physletb.2012.08.020.
154. Gillies, James (23 July 2012). "LHC 2012 proton run extended by seven weeks". CERN bulletin. Retrieved 29 August 2012.
155. "Higgs boson behaving as expected". 3 News NZ. 15 November 2012.
156. Strassler, Matt (2012-11-14). "Higgs Results at Kyoto". Of Particular Significance: Conversations About Science with Theoretical Physicist Matt Strassler. Prof.
157. Sample, Ian (14 November 2012). "Higgs particle looks like a bog Standard Model boson, say scientists". The Guardian. London. Retrieved 15 November 2012.
158. "CERN experiments observe particle consistent with long-sought Higgs boson". CERN press release. 4 July 2012. Retrieved 12 November 2016.
159. "Person Of The Year 2012". Time. 19 December 2012.
160. "Higgs Boson Discovery Has Been Confirmed". Forbes. Retrieved 2013-10-09.
161. Slate Video Staff (2012-09-11). "Higgs Boson Confirmed; CERN Discovery Passes Test". Slate.com. Retrieved 2013-10-09.
162. "The Year Of The Higgs, And Other Tiny Advances In Science". NPR. 2013-01-01. Retrieved 2013-10-09.
163. "Confirmed: the Higgs boson does exist". The Sydney Morning Herald. 4 July 2012.
164. "AP CERN chief: Higgs boson quest could wrap up by midyear". MSNBC. Associated Press. 2013-01-27. Retrieved 20 February 2013. Rolf Heuer, director of [CERN], said he is confident that "towards the middle of the year, we will be there." – Interview by AP, at the World Economic Forum, 26 Jan 2013.
165. Boyle, Alan (2013-02-16). "Will our universe end in a 'big slurp'? Higgs-like particle suggests it might". NBCNews.com – cosmic log. Retrieved 20 February 2013. 'it's going to take another few years' after the collider is restarted to confirm definitively that the newfound particle is the Higgs boson.
166. Gillies, James (2013-03-06). "A question of spin for the new boson". CERN. Retrieved 7 March 2013.

167. Adam Falkowski (writing as 'Jester') (2013-02-27). "When shall we call it Higgs?". Résonaances particle physics blog. Retrieved 7 March 2013.

168. CMS Collaboration (February 2013). "Study of the Mass and Spin-Parity of the Higgs Boson Candidate via Its Decays to Z Boson Pairs". Phys. Rev. Lett. American Physical Society. 110 (8): 081803. arXiv:1212.6639Freely accessible. Bibcode:2013PhRvL.110h1803C. doi:10.1103/PhysRevLett.110.081803. PMID 23473131. Retrieved 15 September 2014.

169. ATLAS Collaboration (7 October 2013). "Evidence for the spin-0 nature of the Higgs boson using ATLAS data". Phys. Lett. B. American Physical Society. 726 (1–3): 120–144. Bibcode:2013PhLB..726..120A. doi:10.1016/j.physletb.2013.08.026. Retrieved 15 September 2014.

170. Chatrchyan, S.; et al. (CMS collaboration) (2013). "Higgs-like Particle in a Mirror". Physical Review Letters. American Physical Society. 110 (8): 081803. Bibcode:2013PhRvL.110h1803C. doi:10.1103/PhysRevLett.110.081803. PMID 23473131. Retrieved 26 February 2013.

171. ATLAS; Collaborations, CMS (2016). "Measurements of the Higgs boson production and decay rates and constraints on its couplings from a combined ATLAS and CMS analysis of the LHC pp collision data at $\sqrt{s}=$ 7 and 8 TeV". arXiv:1606.02266Freely accessible [hep-ex]. line feed character in |title= at position 63 (help)

172. ATLAS collaboration (2016-08-08). "Combined measurements of the Higgs boson production and decay rates in $H \to ZZ* \to 4\ell$ and $H \to \gamma\gamma$ final states using pp collision data at $\sqrt{s}$ = 13 TeV in the ATLAS experiment". Retrieved 2016-08-27.

173. CMS collaboration (2016-08-05). "Updated measurements of Higgs boson production in the diphoton decay channel at $\sqrt{s}$=13 TeV in pp collisions at CMS.". Retrieved 2016-08-27.

174. CMS collaboration (2016-08-05). "Measurements of properties of the Higgs boson and search for an additional resonance in the four-lepton final state at sqrt(s) = 13 TeV". Retrieved 2016-08-27.

175. Liu, G. Z.; Cheng, G. (2002). "Extension of the Anderson-Higgs mechanism". Physical Review B. 65 (13): 132513. arXiv:cond-mat/0106070Freely accessible. Bibcode:2002PhRvB..65m2513L. doi:10.1103/PhysRevB.65.132513.

176. Editorial (2012-03-21). "Mass appeal: As physicists close in on the Higgs boson, they should resist calls to change its name". Nature. 483, 374 (7390): 374. Bibcode:2012Natur.483..374.. doi:10.1038/483374a. Retrieved 21 January 2013.

177. Becker, Kate (2012-03-29). "A Higgs by Any Other Name". "NOVA" (PBS) physics. Retrieved 21 January 2013.

178. "Frequently Asked Questions: The Higgs!". The Bulletin. CERN. Retrieved 18 July 2012.

179. Woit's physics blog "Not Even Wrong": Anderson on Anderson-Higgs 2013-04-13

180. Sample, Ian (2012-07-04). "Higgs boson's many great minds cause a Nobel prize headache". The Guardian. London. Retrieved 23 July 2013.

181. Peskin, M. (July 2012). "40 Years of the Higgs Boson" (PDF). Presentation at SSI 2012. Standford/SSI 2012. pp. 3–5. Retrieved 21 January 2013. quoting Lee's ICHEP 1972 presentation at Fermilab: "...which is known as the Higgs mechanism..." and "Lee's locution" – his footnoted explanation of this shorthand

182. "Rochester's Hagen Sakurai Prize Announcement" (Press release). University of Rochester. 2010.

183. C.R. Hagen Sakurai Prize Talk (YouTube). 2010.

184. Cho, A (2012-09-14). "Particle physics. Why the 'Higgs'?" (PDF). Science. 337 (6100): 1287. doi:10.1126/science.337.6100.1287. PMID 22984044.

185. Weinberg, Steven (2012-05-10). "The Crisis of Big Science". The New York Review of Books. footnote 1. Retrieved 12 February 2013.

186. Examples of early papers using the term "Higgs boson" include 'A phenomenological profile of the Higgs boson' (Ellis, Gaillard and Nanopoulos, 1976), 'Weak interaction theory and neutral currents' (Bjorken, 1977), and 'Mass of the Higgs boson' (Wienberg, received 1975)

187. Leon Lederman; Dick Teresi (2006). The God Particle: If the Universe Is the Answer, What Is the Question?. Houghton Mifflin Harcourt. ISBN 0-547-52462-5.

188. Kelly Dickerson (September 8, 2014). "Stephen Hawking Says 'God Particle' Could Wipe Out the Universe". livescience.com.

189. Jim Baggott (2012). Higgs: The invention and discovery of the 'God Particle'. Oxford University Press. ISBN 978-0-19-165003-1.

190. Scientific American Editors (2012). The Higgs Boson: Searching for the God Particle. Macmillan. ISBN 978-1-4668-2413-3.

191. Ted Jaeckel (2007). The God Particle: The Discovery and Modeling of the Ultimate Prime Particle. Universal-Publishers. ISBN 978-1-58112-959-5.

192. "A Supercompetition For Illinois". Chicago Tribune. 1986-10-31. Retrieved 16 January 2013. The SSC, proposed by the U.S. Department of Energy in 1983, is a mind-bending project ... this gigantic laboratory ... this titanic project

193. Diaz, Jesus (2012-12-15). "This Is [The] World's Largest Super Collider That Never Was". Gizmodo. Retrieved 16 January 2013. ...this titanic complex...

194. Abbott, Charles (June 1987). "Illinois Issues journal, June 1987". p. 18. Lederman,

195. Kevles, Dan. "Good-bye to the SSC: On the Life and Death of the Superconducting Super Collider"

196. Calder, Nigel (2005). Magic Universe:A Grand Tour of Modern Science. pp. 369–370. ISBN 978-0-19-162235-9.

197. Lederman, Leon (1993). The God Particle If the Universe Is the Answer, What Is the Question? (PDF). Dell Publishing. p. Chapter 2, Page 2. ISBN 0-385-31211-3. Retrieved 30 July 2015.

198. Alister McGrath, Higgs boson: the particle of faith, The Daily Telegraph, Published 15 December 2011. Retrieved 15 December 2011.

199. Sample, Ian (3 March 2009). "Father of the God particle: Portrait of Peter Higgs

unveiled". London: The Guardian. Retrieved 24 June 2009.

200. Chivers, Tom (2011-12-13). "How the 'God particle' got its name". The Telegraph. London. Retrieved 2012-12-03.
201. Key scientist sure "God particle" will be found soon Reuters news story. 7 April 2008.
202. "Interview: the man behind the 'God particle'", New Scientist 13 September 2008, pp. 44–5 (original interview in the Guardian: Father of the 'God Particle', June 30, 2008)
203. Sample, Ian (2010). Massive: The Hunt for the God Particle. pp. 148–149 and 278–279. ISBN 978-1-905264-95-7.
204. Cole, K. (2000-12-14). "One Thing Is Perfectly Clear: Nothingness Is Perfect". Los Angeles Times. p. 'Science File'.
205. Sample, Ian (12 June 2009). "Higgs competition: Crack open the bubbly, the God particle is dead". The Guardian. London. Retrieved 4 May 2010.
206. Gordon, Fraser (5 July 2012). "Introducing the higgson". physicsworld.com. Retrieved 25 August 2012.
207. Wolchover, Natalie (2012-07-03). "Higgs Boson Explained: How 'God Particle' Gives Things Mass". Huffington Post. Retrieved 21 January 2013.
208. Oliver, Laura (2012-07-04). "Higgs boson: how would you explain it to a seven-year-old?". The Guardian. London. Retrieved 21 January 2013.
209. Zimmer, Ben (2012-07-15). "Higgs boson metaphors as clear as molasses". The Boston Globe. Retrieved 21 January 2013.
210. "The Higgs particle: an analogy for Physics classroom (section)". www.lhc-closer.es (a collaboration website of LHCb physicist Xabier Vidal and High School Teachers at CERN educator Ramon Manzano). Retrieved 2013-01-09.
211. Flam, Faye (2012-07-12). "Finally – A Higgs Boson Story Anyone Can Understand". The Philadelphia Inquirer (philly.com). Retrieved 21 January 2013.
212. Sample, Ian (2011-04-28). "How will we know when the Higgs particle has been detected?". The Guardian. London. Retrieved 21 January 2013.
213. Miller, David. "A quasi-political Explanation of the Higgs Boson; for Mr Waldegrave, UK Science Minister 1993". Retrieved 10 July 2012.
214. Kathryn Grim. "Ten things you may not know about the Higgs boson". Symmetry Magazine. Retrieved 10 July 2012.
215. David Goldberg, Associate Professor of Physics, Drexel University (2010-10-17). "What's the Matter with the Higgs Boson?". io9.com "Ask a physicist". Retrieved 21 January 2013.
216. The Nobel Prize in Physics 1979 – official Nobel Prize website.
217. The Nobel Prize in Physics 1999 – official Nobel Prize website.
218. – official Nobel Prize website.

219. Daigle, Katy (10 July 2012). "India: Enough about Higgs, let's discuss the boson". AP News. Retrieved 10 July 2012.

220. Bal, Hartosh Singh (19 September 2012). "The Bose in the Boson". New York Times. Retrieved 21 September 2012.

221. Alikhan, Anvar (16 July 2012). "The Spark In A Crowded Field". Outlook India. Retrieved 10 July 2012.

CHAPTER 5

Hadrons

Chapter Contents

A hadron is any particle that is made from quarks, anti-quarks and gluons. Hadrons are viewed as being composed of quarks, either as quark-antiquark pairs (mesons) or as three quarks (baryons). There is much more to the picture than this, however, because the constituent quarks are surrounded by a cloud of gluons, the exchange particles for the color force.

INTRODUCTION

In particle physics, a hadron is a composite particle made of quarks held together by the strong force in a similar way as molecules are held together by the electromagnetic force.

Hadrons are categorized into two families: baryons, made of three quarks, and mesons, made of one quark and one antiquark. Protons and neutrons are examples of baryons; pions are an example of a meson. Hadrons containing more than three valence quarks (exotic hadrons) have been discovered in recent years. A tetraquark state (an exotic meson), named the $Z(4430)^-$, was discovered in 2007 by the Belle Collaboration and confirmed as a resonance in 2014 by the LHCb collaboration. Two pentaquark states (exotic baryons), named P+c(4380) and P+c(4450), were discovered in 2015 by the LHCb collaboration. There are several more exotic hadron candidates, and other color-singlet quark combinations may also exist.

Of the hadrons, protons are stable, and neutrons bound within atomic nuclei are stable. Other hadrons are unstable under ordinary conditions; free neutrons decay with a half-life of about 611 seconds. Experimentally, hadron physics is studied by colliding protons or nuclei of heavy elements such as lead, and detecting the debris in the produced particle showers.

ETYMOLOGY

The term "hadron" was introduced by Lev B. Okun in a plenary talk at the 1962 International Conference on High Energy Physics. In this talk he said:

Notwithstanding the fact that this report deals with weak interactions, we shall frequently have to speak of strongly interacting particles. These particles pose not only numerous scientific problems, but also a terminological problem. The point is that "strongly interacting particles" is a very clumsy term which does not yield itself to the formation of an adjective. For this reason, to take but one instance, decays into strongly interacting particles are called non-leptonic. This definition is not exact because "non-leptonic" may also signify "photonic". In this report I shall call strongly interacting particles "hadrons", and the corresponding decays "hadronic". I hope that this terminology will prove to be convenient. — Lev B. Okun, 1962

PROPERTIES

According to the quark model, the properties of hadrons are primarily determined by their so-called valence quarks. For example, a proton is composed of two up quarks (each with electric charge $+\frac{2}{3}$, for a total of $+\frac{4}{3}$ together) and one down quark (with electric charge $-\frac{1}{3}$). Adding these together yields the proton charge of +1. Although quarks also carry color charge, hadrons must have zero total color charge because of a phenomenon called color confinement. That is, hadrons must be "colorless" or "white". These are the simplest of the two ways: three quarks of different colors, or a quark of one color and an antiquark carrying the corresponding anticolor. Hadrons with the first arrangement are called baryons, and those with the second arrangement are mesons.

Hadrons, however, are not composed of just three or two quarks, because of the strength of the strong force. More accurately, strong force gluons have enough energy (E) to have resonances composed of massive (m) quarks ($E > mc^2$) . Thus, virtual quarks and antiquarks, in a 1:1 ratio, form the majority of massive particles inside a hadron. The two or three quarks that compose a hadron are the excess of quarks vs. antiquarks, and so too in the case of anti-hadrons (anti-particles). Because the virtual quarks are not stable wave packets (quanta), but an irregular and transient phenomena, it is not meaningful to ask which quark is real and which virtual; only the excess is apparent from the outside in the form of a hadron. Massless virtual gluons compose the numerical majority of particles inside hadrons.

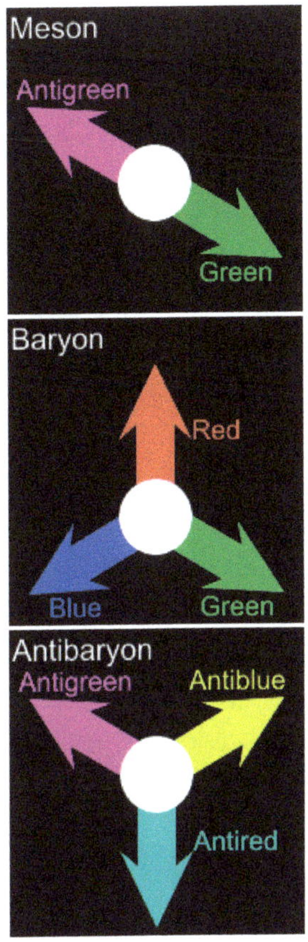

All types of hadrons have zero total color charge. (three examples shown).

Like all subatomic particles, hadrons are assigned quantum numbers corresponding to the representations of the Poincaré group: $J^{PC}(m)$, where J is the spin quantum number, P the intrinsic parity (or P-parity), C the charge conjugation (or C-parity), and m the particle's mass. Note that the mass of a hadron has very little to do with the mass of its valence quarks; rather, due to mass–energy equivalence, most of the mass comes from the large amount of energy associated with the strong interaction. Hadrons may also carry flavor quantum numbers such as isospin (G parity), and strangeness. All quarks carry an additive, conserved quantum number called a baryon number (B), which is $+\frac{1}{3}$ for quarks and $-\frac{1}{3}$ for antiquarks. This means that baryons (groups of three quarks) have B = 1 whereas mesons have B = 0.

Hadrons have excited states known as resonances. Each ground state hadron may have several excited states; several hundreds of resonances have been observed in experiments. Resonances decay extremely quickly (within about 10^{-24} seconds) via the strong nuclear force.

In other phases of matter the hadrons may disappear. For example, at very high temperature and high pressure, unless there are sufficiently many flavors of quarks, the theory of quantum chromodynamics (QCD) predicts that quarks and gluons will no longer be confined within hadrons, "because the strength of the strong interaction diminishes with energy". This property, which is known as asymptotic freedom, has been experimentally confirmed in the energy range between 1 GeV (gigaelectronvolt) and 1 TeV (teraelectronvolt).

All free hadrons except the proton (and antiproton) are unstable.

Baryons

All known baryons are made of three valence quarks, so they are fermions, i.e., they have odd half-integer spin, because they have an odd number of quarks. As quarks possess baryon number B = $\frac{1}{3}$, baryons have baryon number B = 1. The best-known baryons are the proton and the neutron.

One can hypothesise baryons with further quark-antiquark pairs in addition to their three quarks. Hypothetical baryons with one extra quark-antiquark pair (5 quarks in all) are called pentaquarks. As of August 2015, there are two known pentaquarks, P+c(4380) and P+c(4450), both discovered in 2015 by the LHCb collaboration.

Each type of baryon has a corresponding antiparticle (antibaryon) in which quarks are replaced by their corresponding antiquarks. For example, just as a proton is made of two up-quarks and one down-quark, its corresponding antiparticle, the antiproton, is made of two up-antiquarks and one down-antiquark.

MESON

In particle physics, mesons are hadronic subatomic particles composed of one quark and one antiquark, bound together by the strong interaction. Because mesons are composed of quark sub-particles, they have a physical size, with a diameter of roughly one fermi, which is about $\frac{2}{3}$ the size of a proton or neutron. All mesons are unstable, with the longest-lived lasting for only a few hundredths of a microsecond. Charged mesons decay (sometimes through mediating particles) to form electrons and neutrinos. Uncharged mesons may decay to photons. Both of these decays imply that color is no longer a property of the byproducts.

Mesons are not produced by radioactive decay, but appear in nature only as short-lived products of very high-energy interactions in matter, between particles made of quarks, such as the collision of cosmic rays made of nuclei (ordinary protons and neutrons). Mesons are also frequently produced artificially in high-energy particle accelerators in the collisions of protons, anti-protons, or other particles.

In nature, the importance of mesons is that they are the associated quantum-field particles that transmit the nuclear force between hadrons that pull those together into a nucleus. Their effect is analogous to photons that are the force carriers that transmit the electromagnetic force of attraction between oppositely charged protons and electrons that allow individual atoms to exist, and further, to pull atoms together into molecules. Higher energy (more massive) mesons were created momentarily in the Big Bang, but are not thought to play a role in nature today. However, such heavy mesons are regularly created in particle accelerator experiments, in order to understand the nature of the heavier types of quark that compose the heavier mesons.

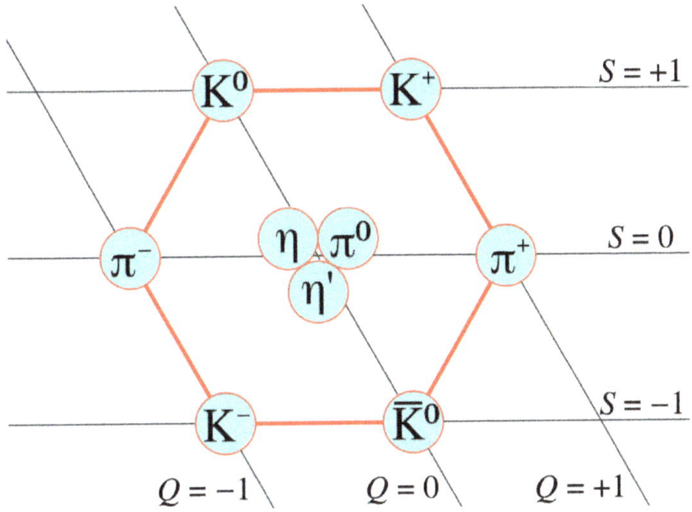

The spin-0 nonet of mesons.

Mesons are part of the hadron particle family, and are defined simply as particles composed of two quarks. The other members of the hadron family are the baryons: subatomic particles composed of three quarks rather than two. Some experiments show evidence of exotic mesons, which do not have the conventional valence quark content of one quark and one antiquark.

Because quarks have a spin of $\frac{1}{2}$, the difference in quark-number between mesons and baryons results in conventional two-quark mesons being bosons, whereas baryons are fermions.

Each type of meson has a corresponding antiparticle (antimeson) in which quarks are replaced by their corresponding antiquarks and vice versa. For example, a positive pion ($\pi+$) is made of one up quark and one down antiquark; and its corresponding antiparticle, the negative pion ($\pi-$), is made of one up antiquark and one down quark.

Because mesons are composed of quarks, they participate in both the weak and strong interactions. Mesons with net electric charge also participate in the electromagnetic interaction. Mesons are classified according to their quark content, total angular momentum, parity and various other properties, such as C-parity and G-parity. Although no meson is stable, those of lower mass are nonetheless more stable than the more massive, and hence are easier to observe and study in particle accelerators or in cosmic ray experiments. Mesons are also typically less massive than baryons, meaning that they are more easily produced in experiments, and thus exhibit certain higher energy phenomena more readily than do baryons composed of the same quarks. For example, the charm quark was first seen in the J/Psi meson (J/ψ) in 1974, and the bottom quark in the upsilon meson (ϒ) in 1977.

History

From theoretical considerations, in 1934 Hideki Yukawa predicted the existence and the approximate mass of the "meson" as the carrier of the nuclear force that holds atomic nuclei together. If there were no nuclear force, all nuclei with two or more protons would fly apart because of the electromagnetic repulsion. Yukawa called his carrier particle the meson, from mesos, the Greek word for "intermediate," because its predicted mass was between that of the electron and that of the proton, which has about 1,836 times the mass of the electron. Yukawa had originally named his particle the "mesotron", but he was corrected by the physicist Werner Heisenberg (whose father was a professor of Greek at the University of Munich). Heisenberg pointed out that there is no "tr" in the Greek word "mesos".

The first candidate for Yukawa's meson, now known in modern terminology as the muon, was discovered in 1936 by Carl David Anderson and others in the decay products of cosmic ray interactions. The mu meson had about the right mass to be Yukawa's carrier of the strong nuclear force, but over the course of the next decade, it became evident that it was not the right particle. It was eventually found that the "mu meson" did not participate in the strong nuclear interaction at all, but rather behaved like a heavy version of the electron, and was eventually classed as a lepton like the electron, rather than a meson. Physicists in making this choice decided that properties other than particle mass should control their classification.

There were years of delays in the subatomic particle research during World War II in 1939–45, with most physicists working in applied projects for wartime necessities. When the war ended in August 1945, many physicists gradually returned to peacetime research. The first true meson to be discovered was what would later be called the "pi meson" (or pion). This discovery was made in 1947, by Cecil Powell, César Lattes, and Giuseppe Occhialini, who were investigating cosmic ray products at the University of Bristol in England, based on photographic films placed in the Andes mountains. Some mesons in these films had about the same mass as the already-known meson, yet seemed to decay into it, leading physicist Robert Marshak to hypothesize in 1947 that it was actually a new and different meson. Over the next few years, more experiments showed that the pion was indeed involved in strong interactions. Unlike other types of force, the pion (as a virtual particle) is also believed to be the primary force carrier for the nuclear force in atomic nuclei. Other mesons, such as the virtual rho mesons are involved in mediating this force as well, but to a lesser extent. Following the discovery of the pion, Yukawa was awarded the 1949 Nobel Prize in Physics for his predictions.

In the past, the word meson was sometimes used to mean any force carrier, such as the "Z^0 meson", which is involved in mediating the weak interaction. However, this spurious usage has fallen out of favor, and mesons are now defined as particles composed of pairs of quarks and antiquarks.

Overview

Spin, Orbital Angular Momentum, and Total Angular Momentum

Spin (quantum number S) is a vector quantity that represents the "intrinsic" angular momentum of a particle. It comes in increments of $\frac{1}{2}$ ħ. The ħ is often dropped because it is the "fundamental" unit of spin, and it is implied that "spin 1" means "spin 1 ħ".(In some systems of natural units, ħ is chosen to be 1, and therefore does not appear in equations.)

Quarks are fermions—specifically in this case, particles having spin $\frac{1}{2}$ (S = $\frac{1}{2}$). Because spin projections vary in increments of 1 (that is 1 ħ), a single quark has a spin vector of length $\frac{1}{2}$, and has two spin projections ($S_z = +\frac{1}{2}$ and $S_z = -\frac{1}{2}$). Two quarks can have their spins aligned, in which case the two spin vectors add to make a vector of length S = 1 and three spin projections ($S_z = +1$, $S_z = 0$, and $S_z = -1$), called the spin-1 triplet. If two quarks have unaligned spins, the spin vectors add up to make a vector of length S = 0 and only one spin projection ($S_z = 0$), called the spin-0 singlet. Because mesons are made of one quark and one antiquark, they can be found in triplet and singlet spin states.

There is another quantity of quantized angular momentum, called the orbital angular momentum (quantum number L), that comes in increments of 1 ħ, which represent the angular momentum due to quarks orbiting around each other. The total angular momentum (quantum number J) of a particle is therefore the combination of intrinsic angular momentum (spin) and orbital angular momentum. It can take any value from J = |L − S| to J = |L + S|, in increments of 1.

Meson angular momentum quantum numbers for *L* = 0, 1, 2, 3				
S	***L***	***J***	***P***	J^P
0	0	0	−	0^-
	1	1	+	1^+
	2	2	−	2^-
	3	3	+	3^+
1	0	1	−	1^-
	1	2, 1, 0	+	$2^+, 1^+, 0^+$
	2	3, 2, 1	−	$3^-, 2^-, 1^-$
	3	4, 3, 2	+	$4^+, 3^+, 2^+$

Particle physicists are most interested in mesons with no orbital angular momentum (L = 0), therefore the two groups of mesons most studied are the S = 1; L = 0 and S = 0; L = 0, which corresponds to J = 1 and J = 0, although they are not the only ones. It is also possible to obtain J = 1 particles from S = 0 and L = 1. How to distinguish between the S = 1, L = 0 and S = 0, L = 1 mesons is an active area of research in meson spectroscopy.

Parity

If the universe were reflected in a mirror, most of the laws of physics would be identical—things would behave the same way regardless of what we call "left" and what we call "right". This concept of mirror reflection is called parity (P). Gravity, the electromagnetic force, and the strong interaction all behave in the same way regardless of whether or not the universe is reflected in a mirror, and thus are said to conserve parity (P-symmetry).

However, the weak interaction does distinguish "left" from "right", a phenomenon called parity violation (P-violation).

Based on this, one might think that, if the wavefunction for each particle (more precisely, the quantum field for each particle type) were simultaneously mirror-reversed, then the new set of wavefunctions would perfectly satisfy the laws of physics (apart from the weak interaction). It turns out that this is not quite true: In order for the equations to be satisfied, the wavefunctions of certain types of particles have to be multiplied by −1, in addition to being mirror-reversed. Such particle types are said to have negative or odd parity (P = −1, or alternatively P = –), whereas the other particles are said to have positive or even parity (P = +1, or alternatively P = +).

For mesons, the parity is related to the orbital angular momentum by the relation:

$$P = (-1)^{L+1}$$

where the L is a result of the parity of the corresponding spherical harmonic of the wavefunction. The '+1' in the exponent comes from the fact that, according to the Dirac equation, a quark and an antiquark have opposite intrinsic parities. Therefore, the intrinsic parity of a meson is the product of the intrinsic parities of the quark (+1) and antiquark (−1). As these are different, their product is −1, and so it contributes a +1 in the exponent.

As a consequence, mesons with no orbital angular momentum (L = 0) all have odd parity (P = −1).

C-Parity

C-parity is only defined for mesons that are their own antiparticle (i.e. neutral mesons). It represents whether or not the wavefunction of the meson remains the same under the interchange of their quark with their antiquark. If

$$|q\bar{q}\rangle = |\bar{q}q\rangle$$

then, the meson is "C even" (C = +1). On the other hand, if

$$|q\bar{q}\rangle = -|\bar{q}q\rangle$$

then the meson is "C odd" (C = −1).

C-parity rarely is studied on its own, but more commonly in combination with P-parity into CP-parity. CP-parity was thought to be conserved, but was later found to be violated in weak interactions.

G-Parity

G–parity is a generalization of the C-parity. Instead of simply comparing the wavefunction after exchanging quarks and antiquarks, it compares the wavefunction after exchanging the meson for the corresponding antimeson, regardless of quark content. In the case of neutral meson, G-parity is equivalent to C-parity because neutral mesons are their own antiparticles.

If

$$|q_1\bar{q_2}\rangle = |\bar{q_1}q_2\rangle$$

then, the meson is “G even” (G = +1). On the other hand, if

$$|q_1\bar{q_2}\rangle = -|\bar{q_1}q_2\rangle$$

then the meson is “G odd” (G = −1).

Isospin and Charge

The concept of isospin was first proposed by Werner Heisenberg in 1932 to explain the similarities between protons and neutrons under the strong interaction. Although they had different electric charges, their masses were so similar that physicists believed that they were actually the same particle. The different electric charges were explained as being the result of some unknown excitation similar to spin. This unknown excitation was later dubbed isospin by Eugene Wigner in 1937. When the first mesons were discovered, they too were seen through the eyes of isospin and so the three pions were believed to be the same particle, but in different isospin states.

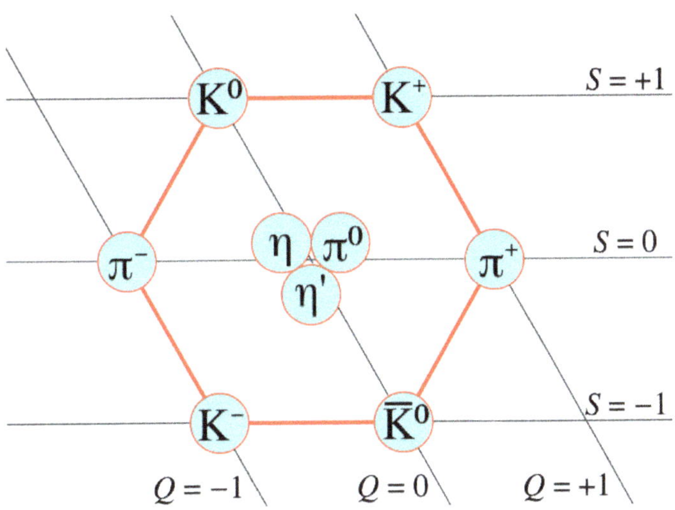

Combinations of one u, d or s quarks and one u, d, or s antiquark in $J^P = 0^-$ configuration form a nonet.

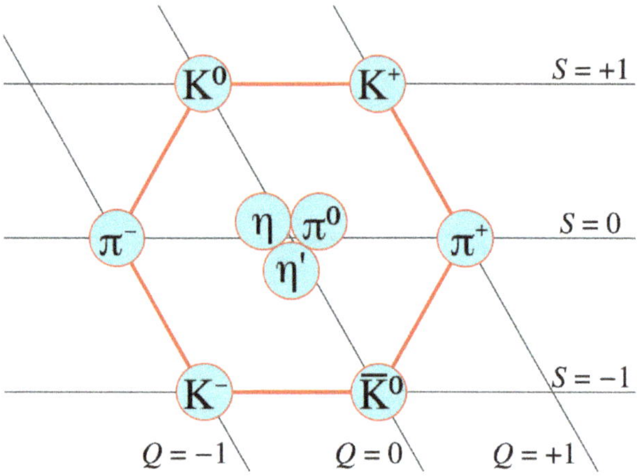

Combinations of one u, d or s quarks and one u, d, or s antiquark in $J^P = 1^-$ configuration also form a nonet.

This belief lasted until Murray Gell-Mann proposed the quark model in 1964 (containing originally only the u, d, and s quarks). The success of the isospin model is now understood to be the result of the similar masses of the u and d quarks. Because the u and d quarks have similar masses, particles made of the same number of them also have similar masses. The exact specific u and d quark composition determines the charge, because u quarks carry charge $+2/3$ whereas d quarks carry charge $-1/3$. For example, the three pions all have different charges (π+ (ud), π0 (a quantum superposition of $u\bar{u}$ and $d\bar{d}$), π− (du)), but have similar masses (c. 140 MeV/c^2) as they are each made of a same number of total of up and down quarks and antiquarks. Under the isospin model, they were considered to be a single particle in different charged states.

The mathematics of isospin was modeled after that of spin. Isospin projections varied in increments of 1 just like those of spin, and to each projection was associated a "charged state". Because the "pion particle" had three "charged states", it was said to be of isospin I = 1. Its "charged states" π+, π0, and π−, corresponded to the isospin projections I_3 = +1, I_3 = 0, and I_3 = −1 respectively. Another example is the "rho particle", also with three charged states. Its "charged states" ρ+, ρ0, and ρ−, corresponded to the isospin projections I_3 = +1, I_3 = 0, and I_3 = −1 respectively. It was later noted that the isospin projections were related to the up and down quark content of particles by the relation

$$I_3 = \frac{1}{2}[(n_u - n_{\bar{u}}) - (n_d - n_{\bar{d}})],$$

where the n's are the number of up and down quarks and antiquarks.

In the "isospin picture", the three pions and three rhos were thought to be the different states of two particles. However, in the quark model, the rhos are excited states of pions. Isospin, although conveying an inaccurate picture of things, is still used to classify hadrons, leading to unnatural and often confusing nomenclature. Because mesons are hadrons, the isospin classification is also used, with $I_3 = +1/2$ for up quarks and down antiquarks, and $I_3 = -1/2$ for up antiquarks and down quarks.

Flavor Quantum Numbers

The strangeness quantum number S (not to be confused with spin) was noticed to go up and down along with particle mass. The higher the mass, the lower the strangeness (the more s quarks). Particles could be described with isospin projections (related to charge) and strangeness (mass). As other quarks were discovered, new quantum numbers were made to have similar description of udc and udb nonets. Because only the u and d mass are similar, this description of particle mass and charge in terms of isospin and flavour quantum numbers only works well for the nonets made of one u, one d and one other quark and breaks down for the other nonets (for example ucb nonet). If the quarks all had the same mass, their behaviour would be called symmetric, because they would all behave in exactly the same way with respect to the strong interaction. However, as quarks do not have the same mass, they do not interact in the same way (exactly like an electron placed in an electric field will accelerate more than a proton placed in the same field because of its lighter mass), and the symmetry is said to be broken.

It was noted that charge (Q) was related to the isospin projection (I_3), the baryon number (B) and flavour quantum numbers (S, C, B′, T) by the Gell-Mann–Nishijima formula:

$$Q = I_3 + \frac{1}{2}(B + S + C + B' + T),$$

where S, C, B′, and T represent the strangeness, charm, bottomness and topness flavour quantum numbers respectively. They are related to the number of strange, charm, bottom, and top quarks and antiquark according to the relations:

$$\begin{aligned} S &= -(n_s - n_{\bar{s}}) \\ C &= +(n_c - n_{\bar{c}}) \\ B' &= -(n_b - n_{\bar{b}}) \\ T &= +(n_t - n_{\bar{t}}), \end{aligned}$$

meaning that the Gell-Mann–Nishijima formula is equivalent to the expression of charge in terms of quark content:

$$Q = \frac{2}{3}[(n_u - n_{\bar{u}}) + (n_c - n_{\bar{c}}) + (n_t - n_{\bar{t}})] - \frac{1}{3}[(n_d - n_{\bar{d}}) + (n_s - n_{\bar{s}}) + (n_b - n_{\bar{b}})].$$

Classification

Mesons are classified into groups according to their isospin (I), total angular momentum (J), parity (P), G-parity (G) or C-parity (C) when applicable, and quark (q) content. The rules for classification are defined by the Particle Data Group, and are rather convoluted. The rules are presented below, in table form for simplicity.

Types of Meson

Mesons are classified into types according to their spin configurations. Some specific configurations are given special names based on the mathematical properties of their spin configuration.

Types of mesons					
Type	***S***	***L***	***P***	***J***	J^P
Pseudoscalar meson	0	0	−	0	0^-
Pseudovector meson	1	1	+	1	1^+
Vector meson	1	0	−	1	1^-
Scalar meson	1	1	+	0	0^+
Tensor meson	1	1	+	2	2^+

Nomenclature

Flavorless Mesons

Flavorless mesons are mesons made of pair of quark and antiquarks of the same flavor (all their flavor quantum numbers are zero: S = 0, C = 0, B′ = 0, T = 0). The rules for flavorless mesons are:

$q\bar{q}$ content	$J^{PC\dagger}$ → / I ↓	$0^{-+}, 2^{-+}, 4^{-+}, \ldots$	$1^{+-}, 3^{+-}, 5^{+-}, \ldots$	$1^{--}, 2^{--}, 3^{--}, \ldots$	$0^{++}, 1^{++}, 2^{++}, \ldots$
$u\bar{d}$ $\frac{u\bar{u}-d\bar{d}}{\sqrt{2}}$ $d\bar{u}$	1	π^+ π^0 π^-	b^+ b^0 b^-	ρ^+ ρ^0 ρ^-	a^+ a^0 a^-
Mix of $u\bar{u}$, $d\bar{d}$, $s\bar{s}$	0	η η'	h h′	ω ϕ	f f′
$c\bar{c}$	0	η_c	h_c	$\psi^{\dagger\dagger}$	χ_c
$b\bar{b}$	0	η_b	h_b	Υ	χ_b
$t\bar{t}$	0	η_t	h_t	θ	χ_t

† The C parity is only relevant to neutral mesons.†† For $J^{PC}=1^{--}$, the ψ is called the J/ψ

In addition:

- When the spectroscopic state of the meson is known, it is added in parentheses.
- When the spectroscopic state is unknown, mass (in MeV/c^2) is added in parentheses.
- When the meson is in its ground state, nothing is added in parentheses.

Flavored Mesons

Flavored mesons are mesons made of pair of quark and antiquarks of different flavors. The rules are simpler in this case: the main symbol depends on the heavier quark, the superscript depends on the charge, and the subscript (if any) depends on the lighter quark. In table form, they are:

antiquark → quark ↓	**up**	**down**	**charm**	**strange**	**top**	**bottom**
up	—	[20]	$\overline{D}^0$	K^+	$\overline{T}^0$	B^+
down	[20]	—	D^-	K^0	T^-	B^0
charm	D^0	D^+	—	D_s^+	$\overline{T}_c^0$	B_c^+
strange	K^-	$\overline{K}^0$	D_s^-	—	T_s^-	B_s^0
top	T^0	T^+	T_c^0	T_s^+	—	T_b^+
bottom	B^-	$\overline{B}^0$	B_c^-	$\overline{B}_s^0$	T_b^-	—

In addition:

- If J^P is in the "normal series" (i.e., $J^P = 0^+, 1^-, 2^+, 3^-, ...$), a superscript ∗ is added.
- If the meson is not pseudoscalar ($J^P = 0^-$) or vector ($J^P = 1^-$), J is added as a subscript.
- When the spectroscopic state of the meson is known, it is added in parentheses.
- When the spectroscopic state is unknown, mass (in MeV/c^2) is added in parentheses.
- When the meson is in its ground state, nothing is added in parentheses.

Exotic Mesons

There is experimental evidence for particles that are hadrons (i.e., are composed of quarks) and are color-neutral with zero baryon number, and thus by conventional definition are mesons. Yet, these particles do not consist of a single quark-antiquark pair, as all the other conventional mesons discussed above do. A tentative category for these particles is exotic mesons.

There are at least five exotic meson resonances that have been experimentally confirmed to exist by two or more independent experiments. The most statistically significant of these is the Z(4430), discovered by the Belle experiment in 2007 and confirmed by LHCb in 2014. It is a candidate for being a tetraquark: a particle composed of two quarks and two antiquarks.

QUARKONIUM

In particle physics, quarkonium (from quark + onium, pl. quarkonia) designates a flavorless meson whose constituents are a quark and its own antiquark. Examples of quarkonia are the J/ψ meson (the ground state of charmonium, cc) and the meson (bottomonium, bb). Because of the high mass of the top quark, toponium does not exist, since the top quark decays through the electroweak interaction before a bound state can form. Usually quarkonium refers only to charmonium and bottomonium, and not to any of the lighter quark–antiquark states. This usage is because the lighter quarks (up, down, and strange) are much less massive than the heavier quarks, and so the physical states actually seen in experiments (η, η′, and π^0 mesons) are quantum mechanical mixtures of the light quark states. The much larger mass differences between the charm and bottom quarks and the lighter quarks results in states that are well defined in terms of a quark–antiquark pair of a given flavor.

Charmonium States

In the following table, the same particle can be named with the spectroscopic notation or with its mass. In some cases excitation series are used: Ψ' is the first excitation of Ψ (for historical reasons, this one is called J/ψ particle); Ψ" is a second excitation, and so on. That is, names in the same cell are synonymous.

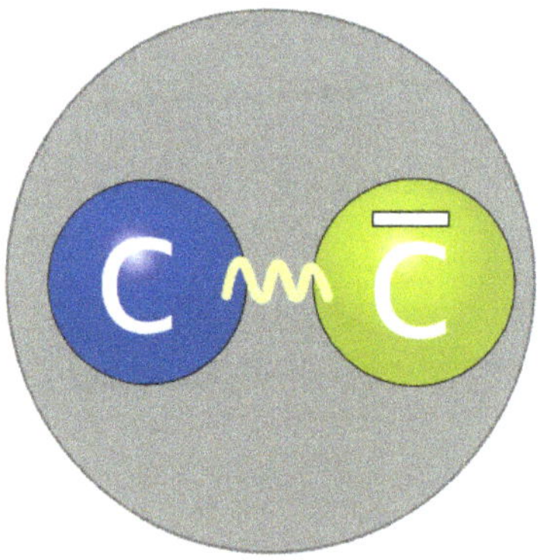

Charmonium

Some of the states are predicted, but have not been identified; others are unconfirmed. The quantum numbers of the X(3872) particle have been measured recently by the LHCb experiment at CERN . This measurement shed some light on its identity, excluding the third option among the three envised, which are :

- a candidate for the 1^1D_2 state;
- a charmonium hybrid state;
- a $D^0\bar{D}^{*0}$ molecule.

In 2005, the BaBar experiment announced the discovery of a new state: Y(4260). CLEO and Belle have since corroborated these observations. At first, Y(4260) was thought to be a charmonium state, but the evidence suggests more exotic explanations, such as a D "molecule", a 4-quark construct, or a hybrid meson.

Term symbol$n^{2S+1}L_J$	$I^G(J^{PC})$	Particle	mass (MeV/c²)
1^1S_0	$0^+(0^{-+})$	$\eta_c(1S)$	2980.3±1.2
1^3S_1	$0^-(1^{--})$	$J/\psi(1S)$	3096.916±0.011
1^1P_1	$0^-(1^{+-})$	$h_c(1P)$	3525.93±0.27
1^3P_0	$0^+(0^{++})$	$\chi_{c0}(1P)$	3414.75±0.31
1^3P_1	$0^+(1^{++})$	$\chi_{c1}(1P)$	3510.66±0.07
1^3P_2	$0^+(2^{++})$	$\chi_{c2}(1P)$	3556.20±0.09
2^1S_0	$0^+(0^{-+})$	$\eta_c(2S)$, or $\eta'c$	3637±4
2^3S_1	$0^-(1^{--})$	$\psi(3686)$	3686.09±0.04
1^1D_2	$0^+(2^{-+})$	$\eta_{c2}(1D)^\dagger$	
1^3D_1	$0^-(1^{--})$	$\psi(3770)$	3772.92±0.35
1^3D_2	$0^-(2^{--})$	$\psi_2(1D)$	
1^3D_3	$0^-(3^{--})$	$\psi_3(1D)^\dagger$	
2^1P_1	$0^-(1^{+-})$	$h_c(2P)^\dagger$	
2^3P_0	$0^+(0^{++})$	$\chi_{c0}(2P)^\dagger$	
2^3P_1	$0^+(1^{++})$	$\chi_{c1}(2P)^\dagger$	
2^3P_2	$0^+(2^{++})$	$\chi_{c2}(2P)^\dagger$	
$?^??_?$	$1^{++}\dagger$	X(3872)	3872.2±0.8
$?^??_?$	$?^?(1^{--})$	Y(4260)	4 2 6 3 + 8 −9

Notes:

* Needs confirmation.

† Predicted, but not yet identified.

† Interpretation as a 1^{--} charmonium state not favored.

Bottomonium States

In the following table, the same particle can be named with the spectroscopic notation or with its mass.

Some of the states are predicted, but have not been identified; others are unconfirmed.

Term symbol $n^{2S+1}L_J$	$I^G(J^{PC})$	Particle	mass (MeV/c²)
1^1S_0	$0^+(0^{-+})$	$\eta_b(1S)$	9390.9±2.8
1^3S_1	$0^-(1^{--})$	$\Upsilon(1S)$	9460.30±0.26
1^1P_1	$0^-(1^{+-})$	$h_b(1P)$	
1^3P_0	$0^+(0^{++})$	$\chi_{b0}(1P)$	9859.44±0.52
1^3P_1	$0^+(1^{++})$	$\chi_{b1}(1P)$	9892.76±0.40
1^3P_2	$0^+(2^{++})$	$\chi_{b2}(1P)$	9912.21±0.40
2^1S_0	$0^+(0^{-+})$	$\eta_b(2S)$	
2^3S_1	$0^-(1^{--})$	$\Upsilon(2S)$	10023.26±0.31
1^1D_2	$0^+(2^{-+})$	$\eta_{b2}(1D)$	
1^3D_1	$0^-(1^{--})$	$\Upsilon(1D)$	
1^3D_2	$0^-(2^{--})$	$\Upsilon_2(1D)$	10161.1±1.7
1^3D_3	$0^-(3^{--})$	$\Upsilon_3(1D)$	
2^1P_1	$0^-(1^{+-})$	$h_b(2P)$	
2^3P_0	$0^+(0^{++})$	$\chi_{b0}(2P)$	10232.5±0.6
2^3P_1	$0^+(1^{++})$	$\chi_{b1}(2P)$	10255.46±0.55
2^3P_2	$0^+(2^{++})$	$\chi_{b2}(2P)$	10268.65±0.55
3^3S_1	$0^-(1^{--})$	$\Upsilon(3S)$	10355.2±0.5
3^3P_J	$0^+(J^{++})$	$\chi_b(3P)$	10530±5 (stat.) ± 9 (syst.)
4^3S_1	$0^-(1^{--})$	$\Upsilon(4S)$ or $\Upsilon(10580)$	10579.4±1.2
5^3S_1	$0^-(1^{--})$	$\Upsilon(5S)$ or $\Upsilon(10860)$	10865±8
6^3S_1	$0^-(1^{--})$	$\Upsilon(11020)$	11019±8

Notes:

* Preliminary results. Confirmation needed.

The χ_b (3P) state was the first particle discovered in the Large Hadron Collider. The article about this discovery was first submitted to arXiv on 21 December 2011. On April 2012, Tevatron's DØ experiment confirms the result in a paper published in Phys. Rev. D.

QCD and Quarkonia

The computation of the properties of mesons in Quantum chromodynamics (QCD) is a fully non-perturbative one. As a result, the only general method available is a direct computation using lattice QCD (LQCD) techniques. However, other techniques are effective for heavy quarkonia as well.

The light quarks in a meson move at relativistic speeds, since the mass of the bound state is much larger than the mass of the quark. However, the speed of the charm and the bottom quarks in their respective quarkonia is sufficiently smaller, so that relativistic effects affect these states much less. It is estimated that the speed, **v**, is roughly 0.3 times the speed of light for charmonia and roughly 0.1 times the speed of light for bottomonia. The computation can then be approximated by an expansion in powers of $\mathbf{v/c}$ and $\mathbf{v^2/c^2}$. This technique is called non-relativistic QCD (NRQCD).

NRQCD has also been quantized as a lattice gauge theory, which provides another technique for LQCD calculations to use. Good agreement with the bottomonium masses has been found, and this provides one of the best non-perturbative tests of LQCD. For charmonium masses the agreement is not as good, but the LQCD community is actively working on improving their techniques. Work is also being done on calculations of such properties as widths of quarkonia states and transition rates between the states.

An early, but still effective, technique uses models of the effective potential to calculate masses of quarkonia states. In this technique, one uses the fact that the motion of the quarks that comprise the quarkonium state is non-relativistic to assume that they move in a static potential, much like non-relativistic models of the hydrogen atom. One of the most popular potential models is the so-called Cornell potential

$$V(r) = -\frac{a}{r} + br$$

where r is the effective radius of the quarkonium state, a and b are parameters. This potential has two parts. The first part, a/r corresponds to the potential induced by one-gluon exchange between the quark and its anti-quark, and is known as the Coulombic part of the potential, since its l/r form is identical to the well-known Coulombic potential induced by the electromagnetic force. The second part, br, is known as the confinement part of the potential, and parameterizes the poorly understood non-perturbative effects of QCD. Generally, when using this approach, a convenient form for the wave function of the quarks is taken, and then a and b are determined by fitting the results of the calculations to the masses of well-measured quarkonium states. Relativistic and other effects can be incorporated into this approach by adding extra terms to the potential, much in the same way that they are for the hydrogen atom in non-relativistic quantum mechanics. This form has been derived from QCD up to $\mathcal{O}(\Lambda_{\rm QCD}^3 r^2)$ by Y. Sumino in 2003. It is popular because it allows for accurate predictions of quarkonia parameters without a lengthy lattice computation, and provides a separation between the short-distance Coulombic effects and the long-distance confinement effects that can be useful in understanding the quark/anti-quark force generated by QCD.

Quarkonia have been suggested as a diagnostic tool of the formation of the quark–gluon plasma: both disappearance and enhancement of their formation depending on the yield of heavy quarks in plasma can occur.

EXOTIC HADRON

Exotic hadrons are subatomic particles composed of quarks and gluons, but which do not have the same quark content as ordinary hadrons: exotic baryons differ from the three-quark (qqq) content of ordinary baryons, and exotic mesons differ from the quark-antiquark (qq) content of ordinary mesons. Hadrons with explicit valence gluon content would also be considered exotic. In theory, there is no limit on the number of quarks in a hadron, as long as the hadron's color charge is white, or color-neutral.

Consistently with ordinary hadrons, exotic hadrons can be classified as being either fermions, like ordinary baryons, or bosons, like ordinary mesons. According to this classification scheme, pentaquarks, containing 5 valence quarks, are exotic baryons, while tetraquarks, containing 4 valence quarks, are exotic mesons. Hexaquarks, containing 6 valence quarks, have not been observed; depending on their quark content, they could be viewed as a dibaryon, or, alternatively, as a triple meson.

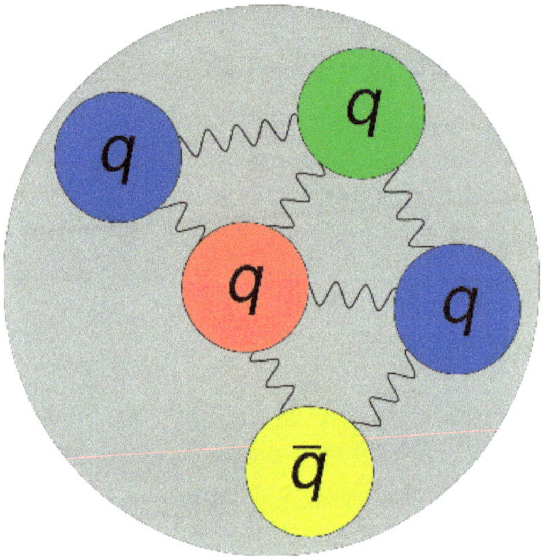

One model of a pentaquark. q indicates a quark, whereas q indicates an antiquark. The wavy lines are gluons, which mediate the strong interaction between the quarks. The colors correspond to the various color charges of quarks. The colors red, green and blue must each be present. The remaining quark and antiquark must share corresponding color and anticolor, here chosen to be blue and antiblue (shown as yellow).

Exotic hadrons can be searched for by looking for S-matrix poles with quantum numbers forbidden to ordinary hadrons. Experimental signatures for such exotic hadrons have been seen by at least 2003 but remain a topic of controversy in particle physics.

Jaffe and Low suggested that the exotic hadrons manifest themselves as poles of the P matrix, and not of the S matrix. Experimental P-matrix poles are determined reliably in both the meson-meson channels and nucleon-nucleon channels.

History

When the quark model was first postulated by Murray Gell-Mann and others in the 1960s, it was to organize the states known then to be in existence in a meaningful way. As quantum chromodynamics (QCD) developed over the next decade, it became apparent that there was no reason why only 3-quark and quark-antiquark combinations could exist. Indeed, Gell-Mann's original 1964 paper alludes to the possibility of exotic hadrons. In addition, it seemed that gluons, the mediator particles of the strong interaction, could also form bound states by themselves (glueballs) and with quarks (hybrid hadrons). Several decades have passed without conclusive evidence of an exotic hadron that could be associated with the S-matrix pole.

In April 2014, the LHCb collaboration confirmed the existence of the $Z(4430)^-$, discovered by Belle, and demonstrated that it must have a minimal quark content of ccdu.

In July 2015, LHCb announced the discovery of two particles, named P+c(4380) and P+c(4450), which must have minimal quark content ccuud, making them pentaquarks.

REFERENCES

1. Choi, S.-K.; Belle Collaboration; et al. (2007). "Observation of a resonance-like structure in the π±Ψ′ mass distribution in exclusive B→Kπ±Ψ′ decays". Physical Review Letters. 100 (14). arXiv:0708.1790Freely accessible. Bibcode:2008PhRvL.100n2001C. doi:10.1103/PhysRevLett.100.142001.
2. LHCb collaboration (2014): Observation of the resonant character of the Z(4430)− state
3. R. Aaij et al. (LHCb collaboration) (2015). "Observation of J/ψp resonances consistent with pentaquark states in Λ0
4. b→J/ψK−p decays". Physical Review Letters. 115 (7). arXiv:1507.03414Freely accessible. Bibcode:2015PhRvL.115g2001A. doi:10.1103/PhysRevLett.115.072001.
5. Lev B. Okun (1962). "The Theory of Weak Interaction". Proceedings of 1962 International Conference on High-Energy Physics at CERN. Geneva. p. 845. Bibcode:1962hep..conf..845O.
6. C. Amsler et al. (Particle Data Group) (2008). "Review of Particle Physics – Quark Model" (PDF). Physics Letters B. 667: 1. Bibcode:2008PhLB..667....1A. doi:10.1016/j.physletb.2008.07.018.
7. S. Bethke (2007). "Experimental tests of asymptotic freedom". Progress in Particle and Nuclear Physics. 58 (2): 351. arXiv:hep-ex/0606035Freely accessible. Bibcode:2007PrPNP..58..351B. doi:10.1016/j.ppnp.2006.06.001.
8. Mysterious Subatomic Particle May Represent Exotic New Form of Matter
9. M.S. Sozzi (2008a). "Parity". Discrete Symmetries and CP Violation: From Experiment to Theory. Oxford University Press. pp. 15–87. ISBN 0-19-929666-9.
10. M.S. Sozzi (2008b). "Charge Conjugation". Discrete Symmetries and CP Violation: From Experiment to Theory. Oxford University Press. pp. 88–120. ISBN 0-19-929666-9.
11. M.S. Sozzi (2008c). "CP-Symmetry". Discrete Symmetries and CP Violation: From Experiment to Theory. Oxford University Press. pp. 231–275. ISBN 0-19-929666-9.

12. C. Amsler (Particle Data Group); et al. (2008). "Review of Particle Physics". Physics Letters B. 667 (1): 1–1340. Bibcode:2008PhLB..667....1A. doi:10.1016/j.physletb.2008.07.018.
13. S.S.M. Wong (1998). "Nucleon Structure". Introductory Nuclear Physics (2nd ed.). New York (NY): John Wiley & Sons. pp. 21–56. ISBN 0-471-23973-9.
14. W.E. Burcham, M. Jobes (1995). Nuclear and Particle Physics (2nd ed.). Longman Publishing. ISBN 0-582-45088-8.
15. R. Shankar (1994). Principles of Quantum Mechanics (2nd ed.). New York (NY): Plenum Press. ISBN 0-306-44790-8.
16. J. Steinberger (1989). "Experiments with high-energy neutrino beams". Reviews of Modern Physics. 61 (3): 533–545. Bibcode:1989RvMP...61..533S. doi:10.1103/RevModPhys.61.533.
17. K. Gottfried, V.F. Weisskopf (1986). "Hadronic Spectroscopy: G-parity". Concepts of Particle Physics. 2. Oxford University Press. pp. 303–311. ISBN 0-19-503393-0.
18. J.W. Cronin (1980). "CP Symmetry Violation—The Search for its origin" (PDF). The Nobel Foundation.
19. V.L. Fitch (1980). "The Discovery of Charge—Conjugation Parity Asymmetry" (PDF). The Nobel Foundation.
20. S.W. Herb; Hom, D.; Lederman, L.; Sens, J.; Snyder, H.; Yoh, J.; Appel, J.; Brown, B.; et al. (1977). "Observation of a Dimuon Resonance at 9.5 Gev in 400-GeV Proton-Nucleus Collisions". Physical Review Letters. 39 (5): 252–255. Bibcode:1977PhRvL..39..252H. doi:10.1103/PhysRevLett.39.252.
21. J.J. Aubert; Becker, U.; Biggs, P.; Burger, J.; Chen, M.; Everhart, G.; Goldhagen, P.; Leong, J.; et al. (1974). "Experimental Observation of a Heavy Particle J". Physical Review Letters. 33 (23): 1404–1406. Bibcode:1974PhRvL..33.1404A. doi:10.1103/PhysRevLett.33.1404.
22. J.E. Augustin; Boyarski, A.; Breidenbach, M.; Bulos, F.; Dakin, J.; Feldman, G.; Fischer, G.; Fryberger, D.; et al. (1974). "Discovery of a Narrow Resonance in e+e− Annihilation". Physical Review Letters. 33 (23): 1406–1408. Bibcode:1974PhRvL..33.1406A. doi:10.1103/PhysRevLett.33.1406.
23. M. Gell-Mann (1964). "A Schematic of Baryons and Mesons". Physics Letters. 8 (3): 214–215. Bibcode:1964PhL.....8..214G. doi:10.1016/S0031-9163(64)92001-3.
24. Ishfaq Ahmad (1965). "the Interactions of 200 MeV $\pi\pm$ -Mesons with Complex Nuclei Proposal to Study the Interactions of 200 MeV $\pi\pm$ -Mesons with Complex Nuclei" (PDF). CERN documents. 3 (5).
25. G. Gamow (1988) [1961]. The Great Physicists from Galileo to Einstein (Reprint ed.). Dover Publications. p. 315. ISBN 978-0-486-25767-9.
26. E. Wigner (1937). "On the Consequences of the Symmetry of the Nuclear Hamiltonian on the Spectroscopy of Nuclei". Physical Review. 51 (2): 106–119. Bibcode:1937PhRv...51..106W. doi:10.1103/PhysRev.51.106.
27. H. Yukawa (1935). "On the Interaction of Elementary Particles" (PDF). Proc. Phys. Math. Soc. Jap. 17 (48).

28. W. Heisenberg (1932). "Über den Bau der Atomkerne I". Zeitschrift für Physik (in German). 77: 1–11. Bibcode:1932ZPhy...77....1H. doi:10.1007/BF01342433.

29. W. Heisenberg (1932). "Über den Bau der Atomkerne II". Zeitschrift für Physik (in German). 78 (3–4): 156–164. Bibcode:1932ZPhy...78..156H. doi:10.1007/BF01337585.

30. W. Heisenberg (1932). "Über den Bau der Atomkerne III". Zeitschrift für Physik (in German). 80 (9–10): 587–596. Bibcode:1933ZPhy...80..587H. doi:10.1007/BF01335696.

31. LHCb collaboration; Aaij, R.; Abellan Beteta, C.; Adeva, B.; Adinolfi, M.; Adrover, C.; Affolder, A.; Ajaltouni, Z.; et al. (February 2013). "Determination of the X(3872) meson quantum numbers". Physical Review Letters. 1302 (22): 6269. arXiv:1302.6269Freely accessible. Bibcode:2013PhRvL.110v2001A. doi:10.1103/PhysRevLett.110.222001.

32. "A new particle discovered by BaBar experiment". Istituto Nazionale di Fisica Nucleare. 6 July 2005. Retrieved 2010-03-06.

33. B. Aubert et al. (BaBar Collaboration) (2005). "Observation of a broad structure in the π+π−J/ψ mass spectrum around 4.26 GeV/c2". Physical Review Letters. 95 (14): 142001. arXiv:hep-ex/0506081Freely accessible. Bibcode:2005PhRvL..95n2001A. doi:10.1103/PhysRevLett.95.142001.

34. ATLAS Collaboration (2012). "Observation of a new χb state in radiative transitions to (1S) and (2S) at ATLAS". arXiv:1112.5154v4Freely accessible [hep-ex].

35. Jonathan Amos (2011-12-22). "LHC reports discovery of its first new particle". BBC.

36. Tevatron experiment confirms LHC discovery of Chi-b (P3) particle

37. Observation of a narrow mass state decaying into Y(1S) + γ in pp collisions at 1.96 TeV

38. Hee Sok Chung; Jungil Lee; Daekyoung Kang (2008). "Cornell Potential Parameters for S-wave Heavy Quarkonia". Journal of the Korean Physical Society. 52 (4): 1151. arXiv:0803.3116Freely accessible. Bibcode:2008JKPS...52.1151C. doi:10.3938/jkps.52.1151.

39. Y. Sumino (2003). "QCD potential as a "Coulomb-plus-linear" potential". Phys. Lett. B. 571: 173–183. arXiv:hep-ph/0303120Freely accessible. Bibcode:2003PhLB..571..173S. doi:10.1016/j.physletb.2003.05.010.

Faddeev–Popov Ghosts

Chapter Contents

In the terminology of theoretical physics, the term "ghosts" is used to identify an object that has no real physical meaning. The name "Faddeev-Popov ghosts" is given to the fictitious fields that were originally introduced in the construction of a manifestly Lorentz covariant quantization of the Yang-Mills field. Later, these objects acquired more widespread application, including in string theory.

INTRODUCTION

In physics, Faddeev–Popov ghosts (also called gauge ghosts or ghost fields) are extraneous fields which are introduced into gauge quantum field theories to maintain the consistency of the path integral formulation. They are named after Ludvig Faddeev and Victor Popov. A more general meaning of the word ghost in theoretical physics is also discussed below.

OVERCOUNTING IN FEYNMAN PATH INTEGRALS

The necessity for Faddeev–Popov ghosts follows from the requirement that quantum field theories yield unambiguous, non-singular solutions. This is not possible in the path integral formulation when a gauge symmetry is present since there is no procedure for selecting among physically equivalent solutions related by gauge transformation. The path integrals overcount field configurations corresponding to the same physical state; the measure of the path integrals contains a factor which does not allow obtaining various results directly from the action.

In particle physics, a ghost condensate is a speculative proposal in which a ghost, an excitation of a field with a wrong sign of the kinetic term, acquires a vacuum expectation value. This phenomenon breaks Lorentz invariance spontaneously. Around the new vacuum state, all excitations have a positive norm, and therefore the probabilities are positive definite.

We have a real scalar field φ with the following action

$$S = \int d^4x \left[aX^2 - bX\right]$$

where a and b are positive constants and

$$X \stackrel{\text{def}}{=} \frac{1}{2}\eta^{\mu\nu}\partial_\mu\phi\partial_\nu\phi$$

using the sign convention in the (+, −, −, −) metric signature.

The theories of ghost condensate predict specific non-Gaussianities of the cosmic microwave background. These theories have been proposed by Nima Arkani-Hamed, Markus Luty, and others.

Unfortunately, this theory allows for superluminal propagation of information in some cases and has no lower bound on its energy. This model doesn't admit a Hamiltonian formulation (the Legendre transform is multi-valued because the momentum function isn't convex) because it is acausal. Quantizing this theory leads to problems.

Faddeev–Popov procedure

It is possible, however, to modify the action, such that methods such as Feynman diagrams will be applicable by adding *ghost fields* which break the gauge symmetry. The ghost fields do not correspond to any real particles in external states: they appear as virtual particles in Feynman diagrams – or as the *absence* of gauge configurations. However, they are a necessary computational tool to preserve unitarity.

The exact form or formulation of ghosts is dependent on the particular gauge chosen, although the same physical results must be obtained with all gauges since the gauge one chooses to carry out calculations is an arbitrary choice. The Feynman-'t Hooft gauge is usually the simplest gauge for this purpose, and is assumed for the rest of this article.

SPIN-STATISTICS RELATION VIOLATED

The Faddeev–Popov ghosts violate the spin-statistics relation, which is another reason why they are often regarded as "non-physical" particles.

For example, in Yang–Mills theories (such as quantum chromodynamics) the ghosts are complex scalar fields (spin 0), but they anti-commute (like fermions).

In general, anti-commuting ghosts are associated with fermionic symmetries, while commuting ghosts are associated with bosonic symmetries.

GAUGE FIELDS AND ASSOCIATED GHOST FIELDS

Every gauge field has an associated ghost, and where the gauge field acquires a mass via the Higgs mechanism, the associated ghost field acquires the same mass (in the Feynman-'t Hooft gauge only, not true for other gauges).

Appearance in Feynman Diagrams

In Feynman diagrams the ghosts appear as closed loops wholly composed of 3-vertices, attached to the rest of the diagram via a gauge particle at each 3-vertex. Their contribution to the S-matrix is exactly cancelled (in the Feynman-'t Hooft gauge) by a contribution from a similar loop of gauge particles with only 3-vertex couplings or gauge attachments to the rest of the diagram. (A loop of gauge particles not wholly composed of 3-vertex couplings is not cancelled by ghosts.) The opposite sign of the contribution of the ghost and gauge loops is due to them having opposite fermionic/bosonic natures. (Closed fermion loops have an extra −1 associated with them; bosonic loops don't.)

GHOST FIELD LAGRANGIAN

The Lagrangian for the ghost fields $c^a(x)$ in Yang–Mills theories (where a is an index in the adjoint representation of the gauge group) is given by

$$\mathcal{L}_{\text{ghost}} = \partial_\mu \bar{c}^a \partial^\mu c^a + g f^{abc} \left(\partial^\mu \bar{c}^a\right) A^b_\mu c^c \ .$$

The first term is a kinetic term like for regular complex scalar fields, and the second term describes the interaction with the gauge fields. Note that in *abelian* gauge theories (such as quantum electrodynamics) the ghosts do not have any effect since $f^{abc} = 0$ and, consequently, the ghost particles do not interact with the gauge fields.

General Ghosts in Theoretical Physics

The Faddeev–Popov ghosts are sometimes referred to as "good ghosts". The "bad ghosts" represent another, more general meaning of the word "ghost" in theoretical physics: states of negative norm—or fields with the wrong sign of the kinetic term, such as Pauli–Villars ghosts—whose existence allows the probabilities to be negative thus violating unitarity.

Changing the Symmetry

Ghost particles could obtain the symmetry or break it in gauge fields. The "good ghost" particles actually obtain the symmetry by unchanging the "gauge fixing lagrangian" in a gauge transformation, while bad ghost particles break the symmetry by bringing in the non-abelian G-matrix which does change the symmetry, and this was the main reason to introduce the Gauge covariant and contravariant derivatives.

REFERENCES

1. Faddeev, L. D.; Popov, V. (1967), "Feynman diagrams for the Yang-Mills field", Phys. Lett. B, 25 (1): 29, Bibcode:1967PhLB...25...29F, doi:10.1016/0370-2693(67)90067-6
2. W. F. Chen. (2008), "Quantum Field Theory and Differential Geometry", Int. J. Geom. Meth. Mod. Phys., 10 (4): 1350003, arXiv:0803.1340v2Freely accessible, doi:10.1142/S0219887813500035
3. Arkani-Hamed, Nima; Cheng, Hsin-Chia; Luty, Markus A.; Mukohyama, Shinji (2004-05-29). "Ghost Condensation and a Consistent Infrared Modification of Gravity". Journal of High Energy Physics. 2004 (05): 074–074. arXiv:hep-th/0312099Freely accessible. Bibcode:2004JHEP...05..074H. doi:10.1088/1126-6708/2004/05/074. ISSN 1029-8479.
4. R. P. Feynman, (1963). "Quantum Theory of Gravitation", Acta Physica Polonica 24, 697–722.

CHAPTER 7

Baryons and Hyperons

Chapter Contents

Particles made up of three quarks are called baryons; the two best known baryons are the proton (made up of two up quarks and one down) and the neutron (two down quarks and one up). Together with the mesons – particles comprised of a quark and an antiquark – baryons form the hadrons (you've heard of hadrons, they're part of the name of the world's most powerful particle collider, the Large Hadron Collider, the LHC).

Hyperons are subatomic particles of the class known as baryons. Like all baryons, they are composed of three quarks. The term hyperon is generally used for a baryon containing one or more strange (s) quarks, as opposed (for example) to the proton and neutron, which contain only up (u) and down (d) quarks. The strange quark being unstable, hyperons decay into lighter baryons (such as protons or neutrons) plus mesons, with typical lifetimes of approximately 1/10 of a nanosecond. At high energies, these lifetimes are sufficient for a hyperon to travel several meters before decaying, since the hyperon can be moving at very nearly the speed of light and thus experience the time-dilation effect of Special Relativity.

INTRODUCTION

A baryon is a composite subatomic particle made up of three quarks (a triquark, as distinct from mesons, which are composed of one quark and one antiquark). Baryons and mesons belong to the hadron family of particles, which are the quark-based particles. The name "baryon" comes from the Greek word for "heavy" (*barys*), because, at the time of their naming, most known elementary particles had lower masses than the baryons.

As quark-based particles, baryons participate in the strong interaction, whereas leptons, which are not quark-based, do not. The most familiar baryons are the protons and neutrons that make up most of the mass of the visible matter in the universe. Electrons (the other major component of the atom) are leptons.

Each baryon has a corresponding antiparticle (antibaryon) where quarks are replaced by their corresponding antiquarks. For example, a proton is made of two up quarks and one down quark; and its corresponding antiparticle, the antiproton, is made of two up antiquarks and one down antiquark.

BACKGROUND

Baryons are strongly interacting fermions, that is, they are acted on by the strong nuclear force and are described by Fermi–Dirac statistics, which apply to all particles obeying the Pauli exclusion principle. This is in contrast to the bosons, which do not obey the exclusion principle.

Baryons, along with mesons, are hadrons, meaning they are particles composed of quarks. Quarks have baryon numbers of $B = 1/3$ and antiquarks have baryon number of $B = -1/3$. The term "baryon" usually refers to *triquarks*—baryons made of three quarks ($B = 1/3 + 1/3 + 1/3 = 1$).

Other exotic baryons have been proposed, such as pentaquarks—baryons made of four quarks and one antiquark ($B = 1/3 + 1/3 + 1/3 + 1/3 - 1/3 = 1$),but their existence is not generally accepted. The particle physics community as a whole did not view their existence as likely in 2006, and in 2008, considered evidence to be overwhelmingly against the existence of the reported pentaquarks. However, in July 2015, the LHCb experiment observed two resonances consistent with pentaquark states in the Λ0b → J/ψK−p decay, with a combined statistical significance of 15σ.

In theory, heptaquarks (5 quarks, 2 antiquarks), nonaquarks (6 quarks, 3 antiquarks), etc. could also exist.

Baryonic Matter

Nearly all matter that may be encountered or experienced in everyday life is baryonic matter, which includes atoms of any sort, and provides those with the property of mass. Non-baryonic matter, as implied by the name, is any sort of matter that is not composed primarily of baryons. This might include neutrinos and free electrons, dark matter, such as supersymmetric particles, axions, and black holes.

The very existence of baryons is also a significant issue in cosmology, because it is assumed that the Big Bang produced a state with equal amounts of baryons and antibaryons. The process by which baryons came to outnumber their antiparticles is called baryogenesis.

Baryogenesis

Experiments are consistent with the number of quarks in the universe being a constant and, to be more specific, the number of baryons being a constant; in technical language, the total baryon number appears to be *conserved.* Within the prevailing Standard Model of particle physics, the number of baryons may change in multiples of three due to the action of sphalerons, although this is rare and has not been observed under experiment. Some grand unified theories of particle physics also predict that a single proton can decay, changing the baryon number by one; however, this has not yet been observed under experiment. The excess of baryons over antibaryons in the present universe is thought to be due to non-conservation of baryon number in the very early universe, though this is not well understood.

Properties

Isospin and Charge

The concept of isospin was first proposed by Werner Heisenberg in 1932 to explain the similarities between protons and neutrons under the strong interaction. Although they had different electric charges, their masses were so similar that physicists believed they were the same particle. The different electric charges were explained as being the result of some unknown excitation similar to spin. This unknown excitation was later dubbed *isospin* by Eugene Wigner in 1937.

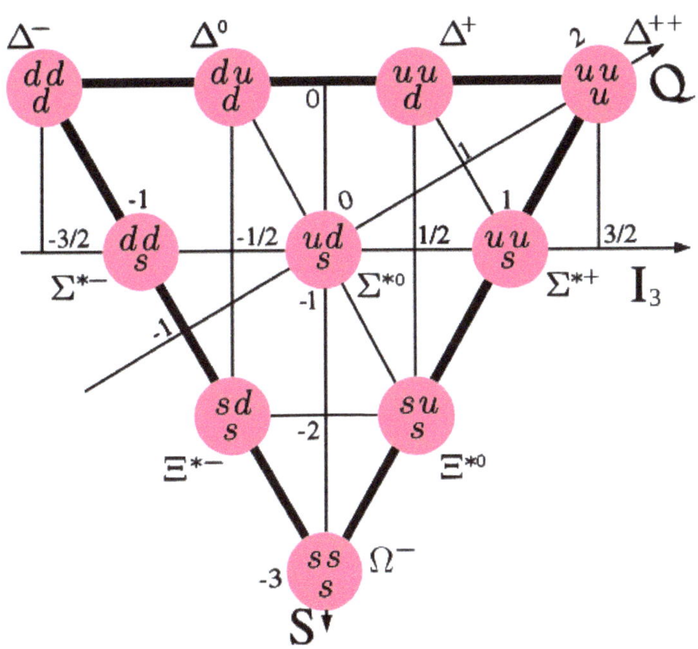

Combinations of three **u, d** or **s** quarks forming baryons with a spin-3/2 form the *uds baryon decuplet*

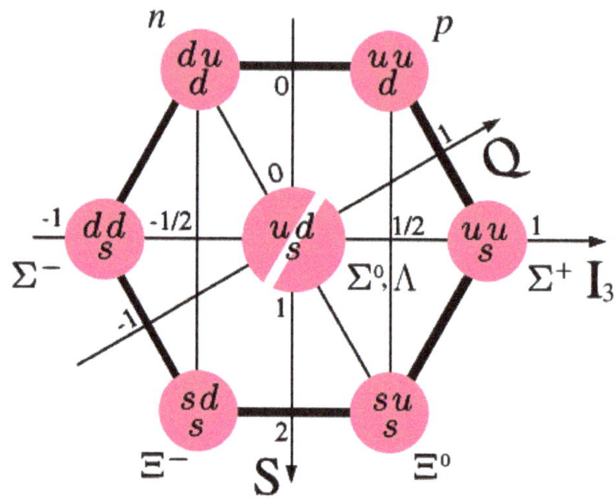

Combinations of three **u, d** or **s** quarks forming baryons with a spin-1/2 form the *uds baryon octet*

This belief lasted until Murray Gell-Mann proposed the quark model in 1964 (containing originally only the u, d, and s quarks). The success of the isospin model is now understood to be the result of the similar masses of the u and d quarks. Since the u and d quarks have similar masses, particles made of the same number then also have similar masses. The exact specific u and d quark composition determines the charge, as u quarks carry charge +2/3 while d quarks carry charge −1/3. For example, the four Deltas all have different charges (Δ^{++} (uuu), Δ^{+} (uud), Δ^{0} (udd), Δ^{-} (ddd)), but have similar masses (~1,232 MeV/c^2) as they are each made of a combination of three u and d quarks. Under the isospin model, they were considered to be a single particle in different charged states.

The mathematics of isospin was modeled after that of spin. Isospin projections varied in increments of 1 just like those of spin, and to each projection was associated a "charged state". Since the "Delta particle" had four "charged states", it was said to be of isospin $I = 3/2$. Its "charged states" Δ++, Δ+, Δ0, and Δ−, corresponded to the isospin projections $I_3 = +3/2$, $I_3 = +1/2$, $I_3 = -1/2$, and $I_3 = -3/2$, respectively. Another example is the "nucleon particle". As there were two nucleon "charged states", it was said to be of isospin 1/2. The positive nucleon N+ (proton) was identified with $I_3 = +1/2$ and the neutral nucleon N0 (neutron) with $I_3 = -1/2$. It was later noted that the isospin projections were related to the up and down quark content of particles by the relation:

$$I_3 = \frac{1}{2}[(n_\mathrm{u} - n_{\bar{\mathrm{u}}}) - (n_\mathrm{d} - n_{\bar{\mathrm{d}}})],$$

where the n's are the number of up and down quarks and antiquarks.

In the "isospin picture", the four Deltas and the two nucleons were thought to be the different states of two particles. However, in the quark model, Deltas are different states of nucleons (the N^{++} or N^{-} are forbidden by Pauli's exclusion principle). Isospin, although conveying an inaccurate picture of things, is still used to classify baryons, leading to unnatural and often confusing nomenclature.

Flavour Quantum Numbers

The strangeness flavour quantum number *S* (not to be confused with spin) was noticed to go up and down along with particle mass. The higher the mass, the lower the strangeness (the more s quarks). Particles could be described with isospin projections (related to charge) and strangeness (mass). As other quarks were discovered, new quantum numbers were made to have similar description of udc and udb octets and decuplets. Since only the u and d mass are similar, this description of particle mass and charge in terms of isospin and flavour quantum numbers works well only for octet and decuplet made of one u, one d, and one other quark, and breaks down for the other octets and decuplets (for example, ucb octet and decuplet). If the quarks all had the same mass, their behaviour would be called *symmetric*, as they would all behave in exactly the same way with respect to the strong interaction. Since quarks do not have the same mass, they do not interact in the same way (exactly like an electron placed in an electric field will accelerate more than a proton placed in the same field because of its lighter mass), and the symmetry is said to be broken.

It was noted that charge (Q) was related to the isospin projection (I_3), the baryon number (B) and flavour quantum numbers (S, C, B', T) by the Gell-Mann–Nishijima formula:

$$Q = I_3 + \frac{1}{2}(B + S + C + B' + T),$$

where S, C, B', and T represent the strangeness, charm, bottomness and topness flavour quantum numbers, respectively. They are related to the number of strange, charm, bottom, and top quarks and antiquark according to the relations:

$$\begin{aligned} S &= -(n_{\mathrm{s}} - n_{\bar{\mathrm{s}}}), \\ C &= +(n_{\mathrm{c}} - n_{\bar{\mathrm{c}}}), \\ B' &= -(n_{\mathrm{b}} - n_{\bar{\mathrm{b}}}), \\ T &= +(n_{\mathrm{t}} - n_{\bar{\mathrm{t}}}), \end{aligned}$$

meaning that the Gell-Mann–Nishijima formula is equivalent to the expression of charge in terms of quark content:

$$Q = \frac{2}{3}[(n_{\mathrm{u}} - n_{\bar{\mathrm{u}}}) + (n_{\mathrm{c}} - n_{\bar{\mathrm{c}}}) + (n_{\mathrm{t}} - n_{\bar{\mathrm{t}}})] - \frac{1}{3}[(n_{\mathrm{d}} - n_{\bar{\mathrm{d}}}) + (n_{\mathrm{s}} - n_{\bar{\mathrm{s}}}) + (n_{\mathrm{b}} - n_{\bar{\mathrm{b}}})].$$

Spin, Orbital Angular Momentum, and Total Angular Momentum

Spin (quantum number *S*) is a vector quantity that represents the "intrinsic" angular momentum of a particle. It comes in increments of 1/2 ħ (pronounced h-bar). The ħ is often dropped because it is the fundamental unit of spin, and it is implied that "spin 1" means "spin 1 ħ". In some systems of natural units, ħ is chosen to be 1, and therefore does not appear anywhere.

Quarks are fermionic particles of spin 1/2 (S = 1/2). Because spin projections vary in increments of 1 (that is 1 ħ), a single quark has a spin vector of length 1/2, and has two spin projections (S_z = +1/2 and S_z = −1/2). Two quarks can have their spins aligned, in which case the two spin vectors add to make a vector of length S = 1 and three spin projections (S_z = +1, S_z = 0, and S_z = −1). If two quarks have unaligned spins, the spin vectors add up to make a vector of length S = 0 and has only one spin projection (S_z = 0), etc. Since baryons are made of three quarks, their spin vectors can add to make a vector of length S = 3/2, which has four spin projections (S_z = +3/2, S_z = +1/2, S_z = −1/2, and S_z = −3/2), or a vector of length S = 1/2 with two spin projections (S_z = +1/2, and S_z = −1/2).

There is another quantity of angular momentum, called the orbital angular momentum, (azimuthal quantum number L), that comes in increments of 1 ħ, which represent the angular moment due to quarks orbiting around each other. The total angular momentum (total angular momentum quantum number J) of a particle is therefore the combination of intrinsic angular momentum (spin) and orbital angular momentum. It can take any value from $J = |L - S|$ to $J = |L + S|$, in increments of 1.

Baryon angular momentum quantum numbers for L = 0, 1, 2, 3				
Spin (S)	**Orbital angular momentum (L)**	**Total angular momentum (J)**	**Parity (P) (See below)**	**Condensed notation (J^P)**
1/2	0	1/2	+	$1/2^+$
	1	3/2, 1/2	–	$3/2^-$, $1/2^-$
	2	5/2, 3/2	+	$5/2^+$, $3/2^+$
	3	7/2, 5/2	–	$7/2^-$, $5/2^-$
3/2	0	3/2	+	$3/2^+$
	1	5/2, 3/2, 1/2	–	$5/2^-$, $3/2^-$, $1/2^-$
	2	7/2, 5/2, 3/2, 1/2	+	$7/2^+$, $5/2^+$, $3/2^+$, $1/2^+$
	3	9/2, 7/2, 5/2, 3/2	–	$9/2^-$, $7/2^-$, $5/2^-$, $3/2^-$

Particle physicists are most interested in baryons with no orbital angular momentum (L = 0), as they correspond to ground states—states of minimal energy. Therefore, the two groups of baryons most studied are the S = 1/2; L = 0 and S = 3/2; L = 0, which corresponds to $J = 1/2^+$ and $J = 3/2^+$, respectively, although they are not the only ones. It is also possible to obtain $J = 3/2^+$ particles from S = 1/2 and L = 2, as well as S = 3/2 and L = 2. This phenomenon of having multiple particles in the same total angular momentum configuration is called *degeneracy*. How to distinguish between these degenerate baryons is an active area of research in baryon spectroscopy.

Parity

This concept of mirror reflection is called *intrinsic parity* or *parity* (P). Gravity, the electromagnetic force, and the strong interaction all behave in the same way regardless of whether or not the universe is reflected in a mirror, and thus are said to conserve parity (P-symmetry). However, the weak interaction *does* distinguish "left" from "right", a phenomenon called parity violation (P-violation).

Based on this, one might think that, if the wavefunction for each particle (in more precise terms, the quantum field for each particle type) were simultaneously mirror-reversed, then the new set of wavefunctions would perfectly satisfy the laws of physics (apart from the weak interaction). It turns out that this is not quite true: In order for the equations to be satisfied, the wavefunctions of certain types of particles have to be multiplied by −1, in addition to being mirror-reversed. Such particle types are said to have *negative* or *odd* parity ($P = -1$, or alternatively $P = -$), while the other particles are said to have *positive* or *even* parity ($P = +1$, or alternatively $P = +$).

For baryons, the parity is related to the orbital angular momentum by the relation:

$$P = (-1)^L.$$

As a consequence, baryons with no orbital angular momentum ($L = 0$) all have even parity ($P = +$).

Nomenclature

Baryons are classified into groups according to their isospin (I) values and quark (q) content. There are six groups of baryons—nucleon (N), Delta (Δ), Lambda (Λ), Sigma (Σ), Xi (Ξ), and Omega (Ω). The rules for classification are defined by the Particle Data Group. These rules consider the up (u), down (d) and strange (s) quarks to be *light* and the charm (c), bottom (b), and top (t) quarks to be *heavy*. The rules cover all the particles that can be made from three of each of the six quarks, even though baryons made of top quarks are not expected to exist because of the top quark's short lifetime. The rules do not cover pentaquarks.

- Baryons with three u and/or d quarks are N's ($I = 1/2$) or Δ's ($I = 3/2$).
- Baryons with two u and/or d quarks are Λ's ($I = 0$) or Σ's ($I = 1$). If the third quark is heavy, its identity is given by a subscript.
- Baryons with one u or d quark are Ξ's ($I = 1/2$). One or two subscripts are used if one or both of the remaining quarks are heavy.
- Baryons with no u or d quarks are Ω's ($I = 0$), and subscripts indicate any heavy quark content.
- Baryons that decay strongly have their masses as part of their names. For example, Σ^0 does not decay strongly, but $\Delta^{++}(1232)$ does.

It is also a widespread (but not universal) practice to follow some additional rules when distinguishing between some states that would otherwise have the same symbol.

- Baryons in total angular momentum $J = 3/2$ configuration that have the same symbols as their $J = 1/2$ counterparts are denoted by an asterisk (*).
- Two baryons can be made of three different quarks in $J = 1/2$ configuration. In this case, a prime (′) is used to distinguish between them.

- *Exception*: When two of the three quarks are one up and one down quark, one baryon is dubbed Λ while the other is dubbed Σ.

Quarks carry charge, so knowing the charge of a particle indirectly gives the quark content. For example, the rules above say that a Λ+c contains a c quark and some combination of two u and/or d quarks. The c quark has a charge of (Q = +2/3), therefore the other two must be a u quark (Q = +2/3), and a d quark (Q = −1/3) to have the correct total charge (Q = +1).

HYPERON

In particle physics, a hyperon is any baryon containing one or more strange quarks, but no charm, bottom, or top quark.

Properties and Behavior of Hyperons

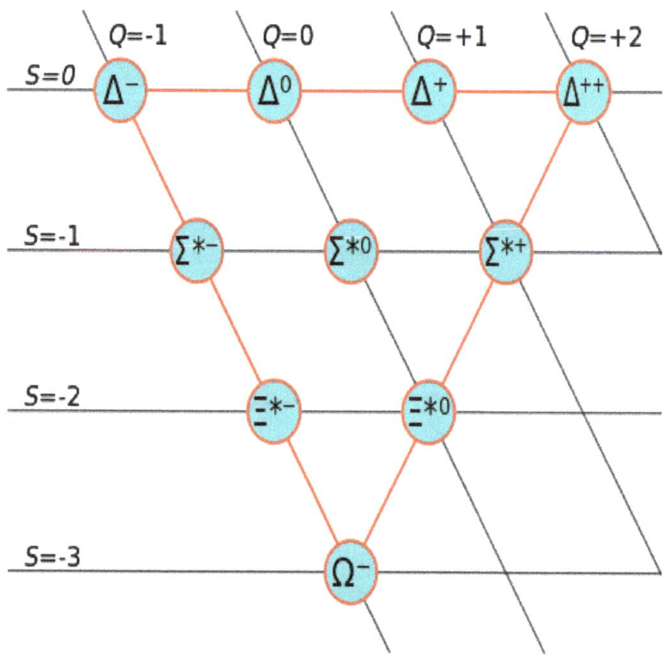

A combination of three u, d or s-quarks with a total spin of 3/2 form the so-called baryon decuplet. The lower six are hyperons.

Being baryons, all hyperons are fermions. That is, they have half-integer spin and obey Fermi–Dirac statistics. They all interact via the strong nuclear force, making them types of hadron. They are composed of three light quarks, at least one of which is a strange quark, which makes them strange baryons. Hyperons decay weakly with non-conserved parity.

List of Hyperons

Particle	Symbol	Makeup	Rest mass MeV/c^2	Isospin I	Spin(Parity) J^P	Q	S	C	B'	Mean lifetime s	Commonly decays to
Lambda [1]	Λ^0	uds	1 115.683(6)	0	$\frac{1}{2}^+$	0	−1	0	0	2.60×10^{-10} [2]	$p^+ + \pi^-$ or $n^0 + \pi^0$
Sigma [3]	Σ^+	uus	1 189.37(0.7)	1	$\frac{1}{2}^+$	+1	−1	0	0	$(8.018 \pm 0.026) \times 10^{-11}$	$p^+ + \pi^0$ or $n^0 + \pi^+$
Sigma [4]	Σ^0	uds	1 192.642(24)	1	$\frac{1}{2}^+$	0	−1	0	0	$(7.4 \pm 0.7) \times 10^{-20}$	$\Lambda^0 + \gamma$
Sigma [5]	Σ^-	dds	1 197.449(30)	1	$\frac{1}{2}^+$	−1	−1	0	0	$(1.479 \pm 0.011) \times 10^{-10}$	$n^0 + \pi^-$
Sigma resonance [6]	$\Sigma^{*+}(1385)$	uus	1 382.8(4)	1	$\frac{3}{2}^+$	+1	−1	0	0		$\Lambda + \pi$ or $\Sigma + \pi$
Sigma resonance [6]	$\Sigma^{*0}(1385)$	uds	1 383.7±1.0	1	$\frac{3}{2}^+$	0	−1	0	0		$\Lambda + \pi$ or $\Sigma + \pi$
Sigma resonance [6]	$\Sigma^{*-}(1385)$	dds	1 387.2(5)	1	$\frac{3}{2}^+$	−1	−1	0	0		$\Lambda + \pi$ or $\Sigma + \pi$
Xi [7]	Ξ^0	uss	1 314.83(20)	$\frac{1}{2}$	$\frac{1}{2}^+$	0	−2	0	0	$(2.90 \pm 0.09) \times 10^{-10}$	$\Lambda^0 + \pi^0$
Xi [8]	Ξ^-	dss	1 321.31(13)	$\frac{1}{2}$	$\frac{1}{2}^+$	−1	−2	0	0	$(1.639 \pm 0.015) \times 10^{-10}$	$\Lambda^0 + \pi^-$
Xi resonance [9]	$\Xi^{*0}(1530)$	uss	1 531.80(32)	$\frac{1}{2}$	$\frac{3}{2}^+$	0	−2	0	0		$\Xi + \pi$
Xi resonance [9]	$\Xi^{*-}(1530)$	dss	1 535.0(6)	$\frac{1}{2}$	$\frac{3}{2}^+$	−1	−2	0	0		$\Xi + \pi$
Omega[10]	Ω^-	sss	1 672.45(29)	0	$\frac{3}{2}^+$	−1	−3	0	0	$(8.21 \pm 0.11) \times 10^{-11}$	$\Lambda^0 + K^-$ or $\Xi^0 + \pi^-$ or $\Xi^- + \pi^0$

Notes:

Since strangeness is conserved by the strong interactions, the ground-state hyperons cannot decay strongly. However, they do participate in strong interactions.

- Λ^0 may also decay on rare occurrences via these processes:

$$\Lambda^0 \rightarrow p^+ + e^- + \bar{\nu}_e$$
$$\Lambda^0 \rightarrow p^+ + \mu^- + \bar{\nu}_\mu$$

- Ξ^0 and Ξ^- are also known as "cascade" hyperons, since they go through a two-step cascading decay into a nucleon.

- The Ω^- has a baryon number of +1 and hypercharge of −2, giving it strangeness of −3.

It takes multiple flavor-changing weak decays for it to decay into a proton or neutron. Murray Gell-Mann's and Yuval Ne'eman's SU(3) model (sometimes called the Eightfold Way) predicted this hyperon's existence, mass and that it will only undergo weak decay processes. Experimental evidence for its existence was discovered in 1964 at Brookhaven National Laboratory. Further examples of its formation and observation using particle accelerators confirmed the SU(3) model.

Hyperon Research

The first research into hyperons happened in the 1950s, and spurred physicists on to the creation of an organized classification of particles. Today, research in this area is carried out on data taken at many facilities around the world, including CERN, Fermilab, SLAC, JLAB, Brookhaven National Laboratory, KEK, and others. Physics topics include searches for CP violation, measurements of spin, studies of excited states (commonly referred to as *spectroscopy*), and hunts for exotic states such as pentaquarks and dibaryons.

REFERENCES

1. C. Amsler et al. (Particle Data Group) (2008). "Review of Particle Physics". Physics Letters B. 667 (1): 1–1340. Bibcode:2008PhLB..667....1A. doi:10.1016/j.physletb.2008.07.018.
2. H. Garcilazo; J. Vijande & A. Valcarce (2007). "Faddeev study of heavy-baryon spectroscopy". Journal of Physics G. 34 (5): 961–976. doi:10.1088/0954-3899/34/5/014.
3. K. Carter (2006). "The rise and fall of the pentaquark". Fermilab and SLAC. Retrieved 2008-05-27.
4. W.-M. Yao et al.(Particle Data Group) (2006). "Review of Particle Physics". Journal of Physics G. 33: 1–1232. arXiv:astro-ph/0601168Freely accessible. Bibcode:2006JPhG...33....1Y. doi:10.1088/0954-3899/33/1/001.
5. D.M. Manley (2005). "Status of baryon spectroscopy". Journal of Physics: Conference Series. 5: 230–237. Bibcode:2005JPhCS...9..230M. doi:10.1088/1742-6596/9/1/043.
6. H. Muir (2003). "Pentaquark discovery confounds sceptics". New Scientist. Retrieved 2008-05-27.
7. S.S.M. Wong (1998a). "Chapter 2—Nucleon Structure". Introductory Nuclear Physics (2nd ed.). New York (NY): John Wiley & Sons. pp. 21–56. ISBN 0-471-23973-9.
8. S.S.M. Wong (1998b). "Chapter 3—The Deuteron". Introductory Nuclear Physics (2nd ed.). New York (NY): John Wiley & Sons. pp. 57–104. ISBN 0-471-23973-9.
9. R. Shankar (1994). Principles of Quantum Mechanics (2nd ed.). New York (NY): Plenum Press. ISBN 0-306-44790-8.
10. E. Wigner (1937). "On the Consequences of the Symmetry of the Nuclear Hamiltonian on the Spectroscopy of Nuclei". Physical Review. 51 (2): 106–119. Bibcode:1937PhRv...51..106W. doi:10.1103/PhysRev.51.106.
11. M. Gell-Mann (1964). "A Schematic of Baryons and Mesons". Physics Letters. 8

(3): 214–215. Bibcode:1964PhL.....8..214G. doi:10.1016/S0031-9163(64)92001-3.

12. W. Heisenberg (1932). "Über den Bau der Atomkerne I". Zeitschrift für Physik (in German). 77: 1–11. Bibcode:1932ZPhy...77....1H. doi:10.1007/BF01342433.
13. W. Heisenberg (1932). "Über den Bau der Atomkerne II". Zeitschrift für Physik (in German). 78 (3–4): 156–164. Bibcode:1932ZPhy...78..156H. doi:10.1007/BF01337585.
14. W. Heisenberg (1932). "Über den Bau der Atomkerne III". Zeitschrift für Physik (in German). 80 (9–10): 587–596. Bibcode:1933ZPhy...80..587H. doi:10.1007/BF01335696.
15. "Particle Data Groups: 2006 Review of Particle Physics – Lambda" (PDF). Retrieved 2008-04-20.
16. "Physics Particle Overview – Baryons". Retrieved 2008-04-20.
17. "Particle Data Groups: 2006 Review of Particle Physics – Sigma+" (PDF). Retrieved 2008-04-20.
18. "Particle Data Groups: 2006 Review of Particle Physics – Sigma0" (PDF). Retrieved 2008-04-20.
19. "Particle Data Groups: 2006 Review of Particle Physics – Sigma-" (PDF). Retrieved 2008-04-20.
20. "Particle Data Groups: 2006 Review of Particle Physics – Sigma(1385)" (PDF). Retrieved 2008-04-20.
21. "Particle Data Groups: 2006 Review of Particle Physics – Xi0" (PDF). Retrieved 2008-04-20.
22. "Particle Data Groups: 2006 Review of Particle Physics – Xi-" (PDF). Retrieved 2008-04-20.
23. "Particle Data Groups: 2006 Review of Particle Physics – Xi(1530)" (PDF). Retrieved 2008-04-20.
24. "Particle Data Groups: 2006 Review of Particle Physics – Omega-" (PDF). Retrieved 2008-04-20.

CHAPTER 8

Phenomena of Quasiparticle

Chapter Contents

- **Introduction**
- **Overview**
- **Davydov Soliton**
- **Dropleton**
- **Exciton**
- **Magnon**
- **Phonon**
- **Polariton**
- **Trion**

Quasiparticle, in physics, a disturbance, in a medium, that behaves as a particle and that may conveniently be regarded as one. A rudimentary analogy is that of a bubble in a glass of beer: the bubble is not really an independent object but a phenomenon, the displacement of a volume of beer by carbon dioxide gas, but, because of the characteristics of the surface of liquid in contact with the gas, the bubble retains a certain identity as it rises and floats. It, like a quasiparticle, carries properties characteristic of objects, such as size, shape, energy, and momentum. Two bubbles can bounce off each other; quasiparticles, too, undergo collisions. Some specific quasiparticles are the exciton, phonon, magnon, and polaron.

INTRODUCTION

In physics, quasiparticles and collective excitations (which are closely related) are emergent phenomena that occur when a microscopically complicated system such as a solid behaves *as if* it contained different weakly interacting particles in free space. For example, as an electron travels through a semiconductor, its motion is disturbed in a complex way by its interactions with all of the other electrons and nuclei; however it *approximately* behaves like an electron with a *different mass* (effective mass) traveling unperturbed through free space. This "electron" with a different mass is called an "electron quasiparticle". In another example, the aggregate motion of electrons in the valence band of a semiconductor is the same as if the semiconductor instead contained positively charged quasiparticles called holes. Other quasiparticles or collective excitations include phonons (particles derived from the vibrations of atoms in a solid), plasmons (particles derived from plasma oscillations), and many others. These particles are typically called "quasiparticles" if they are related to fermions (like electrons and holes), and called "collective excitations" if they are related to bosons (like phonons and plasmons), although the precise distinction is not universally agreed upon.

The quasiparticle concept is most important in condensed matter physics since it is one of the few known ways of simplifying the quantum mechanical many-body problem.

OVERVIEW

Solids are made of only three kinds of particles: Electrons, protons, and neutrons. Quasiparticles are none of these; instead, they are an *emergent phenomenon* that occurs inside the solid. Therefore, while it is quite possible to have a single particle (electron or proton or neutron) floating in space, a *quasi*particle can instead only exist inside the solid.

Motion in a solid is extremely complicated: Each electron and proton is pushed and pulled (by Coulomb's law) by all the other electrons and protons in the solid (which may themselves be in motion). It is these strong interactions that make it very difficult to predict and understand the behavior of solids. On the other hand, the motion of a *non-interacting* particle is quite simple: In classical mechanics, it would move in a straight line, and in quantum mechanics, it would move in a superposition of plane waves. This is the motivation for the concept of quasiparticles: The complicated motion of the *actual* particles in a solid can be mathematically transformed into the much simpler motion of imagined *quasi*particles, which behave more like non-interacting particles.

In summary, quasiparticles are a mathematical tool for simplifying the description of solids. They are not "real" particles inside the solid. Instead, saying "A quasiparticle is present" or "A quasiparticle is moving" is shorthand for saying "A large number of electrons and nuclei are moving in a specific coordinated way."

Relation to Many-Body Quantum Mechanics

The principal motivation for quasiparticles is that it is almost impossible to *directly* describe every particle in a macroscopic system. For example, a barely-visible (0.1mm) grain of sand contains around 10^{17} atoms and 10^{18} electrons. Each of these attracts or repels every other by Coulomb's law. In quantum mechanics, a system is described by a wavefunction, which, if the particles are interacting (as they are in our case), depends on the position of every particle in the system. So, each particle adds three independent variables to the wavefunction, one for each coordinate needed to describe the position of that particle. Because of this, directly approaching the many-body problem of 10^{18} interacting electrons by straightforwardly trying to solve the appropriate Schrödinger equation is impossible in practice, since it amounts to solving a partial differential equation not just in three dimensions, but in 3×10^{18} dimensions – one for each component of the position of each particle.

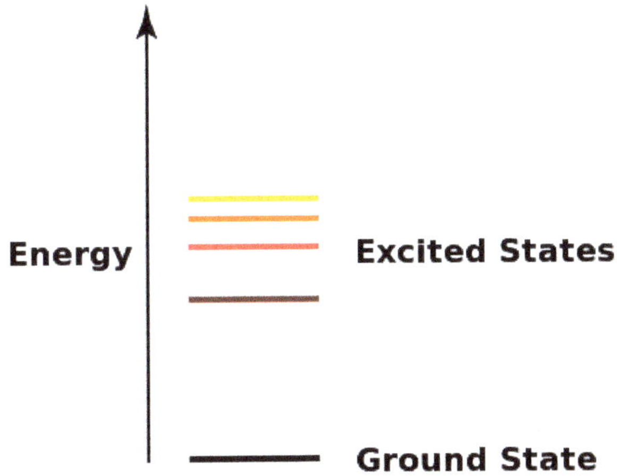

Any system, no matter how complicated, has a ground state along with an infinite series of higher-energy excited states.

One simplifying factor is that the system as a whole, like any quantum system, has a ground state and various excited states with higher and higher energy above the ground state. In many contexts, only the "low-lying" excited states, with energy reasonably close to the ground state, are relevant. This occurs because of the Boltzmann distribution, which implies that very-high-energy thermal fluctuations are unlikely to occur at any given temperature.

Quasiparticles and collective excitations are a type of low-lying excited state. For example, a crystal at absolute zero is in the ground state, but if one phonon is added to the crystal (in other words, if the crystal is made to vibrate slightly at a particular frequency) then the crystal is now in a low-lying excited state. The single phonon is called an *elementary excitation*. More generally, low-lying excited states may contain any number of elementary excitations (for example, many phonons, along with other quasiparticles and collective excitations).

When the material is characterized as having "several elementary excitations", this statement presupposes that the different excitations can be combined together. In other words, it presupposes that the excitations can coexist simultaneously and independently. This is never *exactly* true. For example, a solid with two identical phonons does not have exactly twice the excitation energy of a solid with just one phonon, because the crystal vibration is slightly anharmonic. However, in many materials, the elementary excitations are very *close* to being independent. Therefore, as a *starting point*, they are treated as free, independent entities, and then corrections are included via interactions between the elementary excitations, such as "phonon-phonon scattering".

Therefore, using quasiparticles / collective excitations, instead of analyzing 10^{18} particles, one needs to deal with only a handful of somewhat-independent elementary excitations. It is, therefore, a very effective approach to simplify the many-body problem in quantum mechanics. This approach is not useful for *all* systems, however: In strongly correlated materials, the elementary excitations are so far from being independent that it is not even useful as a starting point to treat them as independent.

Distinction between Quasiparticles and Collective Excitations

Usually, an elementary excitation is called a "quasiparticle" if it is a fermion and a "collective excitation" if it is a boson. However, the precise distinction is not universally agreed upon.

There is a difference in the way that quasiparticles and collective excitations are intuitively envisioned. A quasiparticle is usually thought of as being like a dressed particle: It is built around a real particle at its "core", but the behavior of the particle is affected by the environment. A standard example is the "electron quasiparticle": A real electron particle, in a crystal, behaves as if it had a different mass. On the other hand, a collective excitation is usually imagined to be a reflection of the aggregate behavior of the system, with no single real particle at its "core". A standard example is the phonon, which characterizes the vibrational motion of every atom in the crystal.

However, these two visualizations leave some ambiguity. For example, a magnon in a ferromagnet can be considered in one of two perfectly equivalent ways: (a) as a mobile defect (a misdirected spin) in a perfect alignment of magnetic moments or (b) as a quantum of a collective spin wave that involves the precession of many spins. In the first case, the magnon is envisioned as a quasiparticle, in the second case, as a collective excitation. However, both (a) and (b) are equivalent and correct descriptions. As this example shows, the intuitive distinction between a quasiparticle and a collective excitation is not particularly important or fundamental.

The problems arising from the collective nature of quasiparticles have also been discussed within the philosophy of science, notably in relation to the identity conditions of quasiparticles and whether they should be considered "real" by the standards of, for example, entity realism.

Effect on Bulk Properties

By investigating the properties of individual quasiparticles, it is possible to obtain a great deal of information about low-energy systems, including the flow properties and heat capacity.

In the heat capacity example, a crystal can store energy by forming phonons, and/or forming excitons, and/or forming plasmons, etc. Each of these is a separate contribution to the overall heat capacity.

History

The idea of quasiparticles originated in Lev Landau's theory of Fermi liquids, which was originally invented for studying liquid helium-3. For these systems a strong similarity exists between the notion of quasiparticle and dressed particles in quantum field theory.

The dynamics of Landau's theory is defined by a kinetic equation of the mean-field type. A similar equation, the Vlasov equation, is valid for a plasma in the so-called plasma approximation. In the plasma approximation, charged particles are considered to be moving in the electromagnetic field collectively generated by all other particles, and hard collisions between the charged particles are neglected. When a kinetic equation of the mean-field type is a valid first-order description of a system, second-order corrections determine the entropy production, and generally take the form of a Boltzmann-type collision term, in which figure only "far collisions" between virtual particles. In other words, every type of mean-field kinetic equation, and in fact every mean-field theory, involves a quasiparticle concept.

Examples of Quasiparticles and Collective Excitations

This section contains examples of quasiparticles and collective excitations. The first subsection below contains common ones that occur in a wide variety of materials under ordinary conditions; the second subsection contains examples that arise only in special contexts.

More Common Examples

- In solids, an electron quasiparticle is an electron as affected by the other forces and interactions in the solid. The electron quasiparticle has the same charge and spin as a "normal" (elementary particle) electron, and like a normal electron, it is a fermion. However, its mass can differ substantially from that of a normal electron. Its electric field is also modified, as a result of electric field screening. In many other respects, especially in metals under ordinary conditions, these so-called Landau quasiparticles closely resemble familiar electrons; as Crommie's "quantum corral" showed, an STM can clearly image their interference upon scattering.
- A hole is a quasiparticle consisting of the lack of an electron in a state; it is most commonly used in the context of empty states in the valence band of a semiconductor. A hole has the opposite charge of an electron.
- A phonon is a collective excitation associated with the vibration of atoms in a rigid crystal structure. It is a quantum of a sound wave.
- A magnon is a collective excitation associated with the electrons' spin structure in a crystal lattice. It is a quantum of a spin wave.
- A roton is a collective excitation associated with the rotation of a fluid (often a superfluid). It is a quantum of a vortex.
- In materials, a photon quasiparticle is a photon as affected by its interactions with the material. In particular, the photon quasiparticle has a modified relation between wavelength and energy (dispersion relation), as described by the material's index of refraction. It may also be termed a polariton, especially near a resonance of the material. For example, an exciton-polariton is a superposition of an exciton and a photon; a phonon-polariton is a superposition of a phonon and a photon.

- A plasmon is a collective excitation, which is the quantum of plasma oscillations (wherein all the electrons simultaneously oscillate with respect to all the ions).
- A polaron is a quasiparticle which comes about when an electron interacts with the polarization of its surrounding ions.
- An exciton is an electron and hole bound together.
- A plasmariton is a coupled optical phonon and dressed photon consisting of a plasmon and photon.

More Specialized Examples

- Composite fermions arise in a two-dimensional system subject to a large magnetic field, most famously those systems that exhibit the fractional quantum Hall effect. These quasiparticles are quite unlike normal particles in two ways. First, their charge can be less than the electron charge *e*. In fact, they have been observed with charges of e/3, e/4, e/5, and e/7. Second, they can be anyons, an exotic type of particle that is neither a fermion nor boson.
- Stoner excitations in ferromagnetic metals
- Bogoliubov quasiparticles in superconductors. Superconductivity is carried by Cooper pairs—usually described as pairs of electrons—that move through the crystal lattice without resistance. A broken Cooper pair is called a Bogoliubov quasiparticle. It differs from the conventional quasiparticle in metal because it combines the properties of a negatively charged electron and a positively charged hole (an electron void). Physical objects like impurity atoms, from which quasiparticles scatter in an ordinary metal, only weakly affect the energy of a Cooper pair in a conventional superconductor. In conventional superconductors, interference between Bogoliubov quasiparticles is tough for an STM to see. Because of their complex global electronic structures, however, high-Tc cuprate superconductors are another matter. Thus Davis and his colleagues were able to resolve distinctive patterns of quasiparticle interference in Bi-2212.
- A Majorana fermion is a particle which equals its own antiparticle, and can emerge as a quasiparticle in certain superconductors, or in a quantum spin liquid.
- Magnetic monopoles arise in condensed matter systems such as spin ice and carry an effective magnetic charge as well as being endowed with other typical quasiparticle properties such as an effective mass. They may be formed through spin flips in frustrated pyrochlore ferromagnets and interact through a Coulomb potential.
- Skyrmions
- Spinon is represented by quasiparticle produced as a result of electron spin-charge separation, and can form both quantum spin liquid and strongly correlated quantum spin liquid in some minerals like Herbertsmithite.

DAVYDOV SOLITON

Davydov soliton is a quantum quasiparticle representing an excitation propagating along the protein α-helix self-trapped amide I. It is a solution of the Davydov Hamiltonian. It is named for the Soviet and Ukrainian physicist Alexander Davydov. The Davydov model describes the interaction of the amide I vibrations with the hydrogen bonds that stabilize the α-helix of proteins. The elementary excitations within the α-helix are given by the phonons which correspond to the deformational oscillations of the lattice, and the excitons which describe the internal amide I excitations of the peptide groups. Referring to the atomic structure of an α-helix region of protein the mechanism that creates the Davydov soliton (polaron, exciton) can be described as follows: vibrational energy of the C=O stretching (or amide I) oscillators that is localized on the α-helix acts through a phonon coupling effect to distort the structure of the α-helix, while the helical distortion reacts again through phonon coupling to trap the amide I oscillation energy and prevent its dispersion. This effect is called *self-localization* or *self-trapping*. Solitons in which the energy is distributed in a fashion preserving the helicalsymmetry are dynamically unstable, and such symmetrical solitons once formed decay rapidly when they propagate. On the other hand, an asymmetric soliton which spontaneously breaks the local translational and helical symmetries possesses the lowest energy and is a robust localized entity.

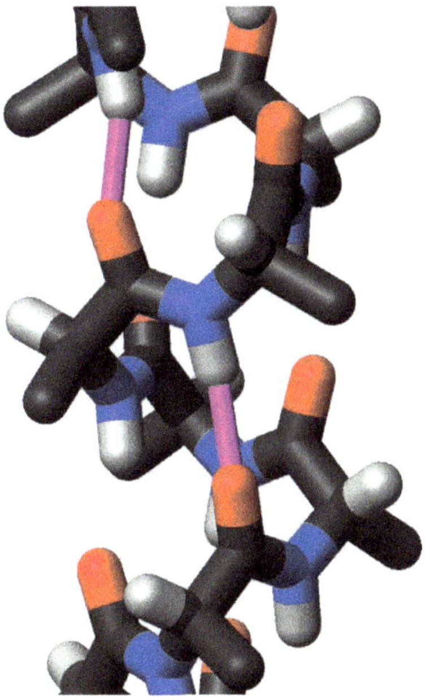

Side view of an α-helix of alanine residues in atomic detail. Protein α-helices provide substrate for Davydov soliton creation and propagation.

DROPLETON

A dropleton or quantum droplet is a quasiparticle comprising a collection of electrons and holes inside a semiconductor. Dropletons give the first known quasiparticle characterization where the quasiparticle behaves like a liquid. The creation of dropletons was announced on 26 February 2014 in a *Nature* article, which presented evidence for the creation of dropletons in an electron–hole plasma inside a gallium arsenide quantum well by ultrashort laser pulses. Their existence was not predicted before the experiment.

Despite the relatively short lifetime of about 25 picoseconds, the dropletons are stable enough to be studied and possess favorable properties for certain investigations of quantum mechanics. They are approximately 200 nanometers wide, the size of the smallest bacteria, for which reason the discoverers have expressed hope that they might one day actually see quantum droplets.

EXCITON

An exciton is a bound state of an electron and an electron hole which are attracted to each other by the electrostatic Coulomb force. It is an electrically neutral quasiparticle that exists in insulators, semiconductors and in some liquids. The exciton is regarded as an elementary excitation of condensed matter that can transport energy without transporting net electric charge.

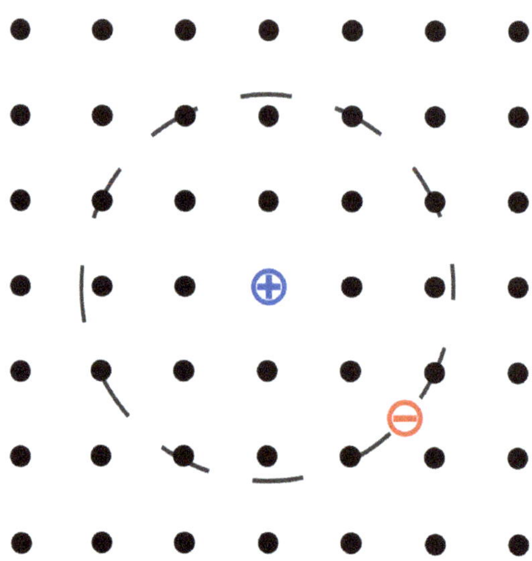

Frenkel exciton, bound electron-hole pair where the hole is localized at a position in the crystal represented by black dots

An exciton can form when a photon is absorbed by a semiconductor. This excites an electron from the valence band into the conduction band. In turn, this leaves behind a positively charged electron hole (an abstraction for the location from which an electron was moved). The electron in the conduction band is then effectively attracted to this localized hole by the repulsive Coulomb forces from large numbers of electrons surrounding the hole and excited electron. This attraction provides a stabilizing energy balance. Consequently, the exciton has slightly less energy than the unbound electron and hole. The wavefunction of the bound state is said to be *hydrogenic*, an exotic atom state akin to that of a hydrogen atom. However, the binding energy is much smaller and the particle's size much larger than a hydrogen atom. This is because of both the screening of the Coulomb force by other electrons in the semiconductor (i.e., its dielectric constant), and the small effective masses of the excited electron and hole. The recombination of the electron and hole, i.e. the decay of the exciton, is limited by resonance stabilization due to the overlap of the electron and hole wave functions, resulting in an extended lifetime for the exciton.

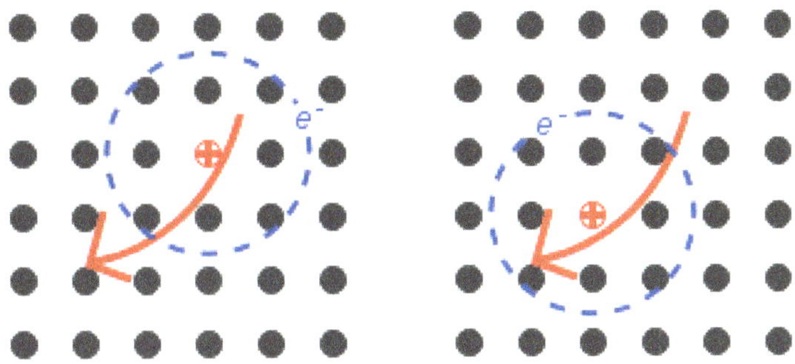

Wannier-Mott exciton, bound electron-hole pair that is not localized at a crystal position. This figure schematically shows diffusion of the exciton across the lattice.

The electron and hole may have either parallel or anti-parallel spins. The spins are coupled by the exchange interaction, giving rise to exciton fine structure. In periodic lattices, the properties of an exciton show momentum (k-vector) dependence.

The concept of excitons was first proposed by Yakov Frenkel in 1931, when he described the excitation of atoms in a lattice of insulators. He proposed that this excited state would be able to travel in a particle-like fashion through the lattice without the net transfer of charge.

Classification

Excitons may be treated in two limiting cases, depending on the properties of the material in question.

Frenkel Excitons

In materials with a small dielectric constant, the Coulomb interaction between an electron and a hole may be strong and the excitons thus tend to be small, of the same order as the size of the unit cell. Molecular excitons may even be entirely located on the same molecule, as in fullerenes. This *Frenkel exciton*, named after Yakov Frenkel, has a typical binding energy on the order of 0.1 to 1 eV. Frenkel excitons are typically found in alkali halide crystals and in organic molecular crystals composed of aromatic molecules, such as anthracene and tetracene.

Wannier-Mott Excitons

In semiconductors, the dielectric constant is generally large. Consequently, electric field screening tends to reduce the Coulomb interaction between electrons and holes. The result is a *Wannier exciton*, which has a radius larger than the lattice spacing. Small effective mass of electrons that is typical of semiconductors also favors large exciton radii. As a result, the effect of the lattice potential can be incorporated into the effective masses of the electron and hole. Likewise, because of the lower masses and the screened Coulomb interaction, the binding energy is usually much less than that of a hydrogen atom, typically on the order of 0.01eV. This type of exciton was named for Gregory Wannier and Nevill Francis Mott. Wannier-Mott excitons are typically found in semiconductor crystals with small energy gaps and high dielectric constants, but have also been identified in liquids, such as liquid xenon. They are also known as *large excitons*.

In single-wall carbon nanotubes, excitons have both Wannier-Mott and Frenkel character. This is due to the nature of the Coulomb interaction between electrons and holes in one-dimension. The dielectric function of the nanotube itself is large enough to allow for the spatial extent of the wave function to extend over a few to several nanometers along the tube axis, while poor screening in the vacuum or dielectric environment outside of the nanotube allows for large (0.4 to 1.0eV) binding energies.

Often there is more than one band to choose from for the electron and the hole leading to different types of excitons in the same material. Even high-lying bands can be effective as femtosecond two-photon experiments have shown. At cryogenic temperatures, many higher excitonic levels can be observed approaching the edge of the band, forming a series of spectral absorption lines that are in principle similar to hydrogen spectral series.

Charge-Transfer Excitons

An intermediate case between Frenkel and Wannier excitons, *charge-transfer excitons* (sometimes called simply *CT excitons*) occur when the electron and the hole occupy adjacent molecules. They occur primarily in ionic crystals. Unlike Frenkel and Wannier excitons they display a static electric dipole moment.

Surface Excitons

At surfaces it is possible for so called *image states* to occur, where the hole is inside the solid and the electron is in the vacuum. These electron-hole pairs can only move along the surface.

Atomic and Molecular Excitons

Alternatively, an exciton may be an excited state of an atom, ion, or molecule, the excitation wandering from one cell of the lattice to another.

When a molecule absorbs a quantum of energy that corresponds to a transition from one molecular orbital to another molecular orbital, the resulting electronic excited state is also properly described as an exciton. An electron is said to be found in the lowest unoccupied orbital and an electron hole in the highest occupied molecular orbital, and since they are found within the same molecular orbital manifold, the electron-hole state is said to be bound. Molecular excitons typically have characteristic lifetimes on the order of nanoseconds, after which the ground electronic state is restored and the molecule undergoes photon or phonon emission. Molecular excitons have several interesting properties, one of which is energy transfer whereby if a molecular exciton has proper energetic matching to a second molecule's spectral absorbance, then an exciton may transfer (*hop*) from one molecule to another. The process is strongly dependent on intermolecular distance between the species in solution, and so the process has found application in sensing and *molecular rulers*.

The hallmark of molecular excitons in organic molecular crystals are doublets and/or triplets of exciton absorption bands strongly polarized along crystallographic axes. In these crystals an elementary cell includes several molecules sitting in symmetrically identical positions, which results in the level degeneracy that is lifted by intermolecular interaction. As a result, absorption bands are polarized along the symmetry axes of the crystal. Such multiplets were discovered by Antonina Prikhot'ko and their genesis was proposed by Alexander Davydov. It is known as 'Davydov splitting'.

Giant Oscillator Strength of Bound Excitons

Excitons are lowest excited states of the electronic subsystem of pure crystals. Impurities can bind excitons, and when the bound state is shallow, the oscillator strength for producing bound excitons is so high that impurity absorption can compete with intrinsic exciton absorption even at rather low impurity concentrations. This phenomenon is generic and applicable both to the large radius (Wannier-Mott) excitons and molecular (Frenkel) excitons. Hence, excitons bound to impurities and defects possess giant oscillator strength.

Self-Trapping of Excitons

In crystals excitons interact with phonons, the lattice vibrations. If this coupling is weak as in typical semiconductors such as GaAs or Si, excitons are scattered by phonons. However, when the coupling is strong, excitons can be self-trapped. Self-trapping results in dressing excitons with a dense cloud of virtual phonons which strongly suppresses the ability of excitons to move across the crystal. In simpler terms, this means a local deformation of the crystal lattice around the exciton. Self-trapping can be achieved only if the energy of this deformation can compete with the width of the exciton band. Hence, it should be of atomic scale, of about an electron-volt.

Self-trapping of excitons is similar to forming strong-coupling polarons but with three essential differences. First, self-trapped exciton states are always of a small radius, of the order of lattice constant, due to their electric neutrality. Second, there exists a self-trapping barrier separating free and self-trapped states, hence, free excitons are metastable. Third, this barrier enables coexistence of free and self-trapped states of excitons. This means that spectral lines of free excitons and wide bands of self-trapped excitons can be seen simultaneously in absorption and luminescence spectra. It is interesting that while the self-trapped states are of lattice-spacing scale, the barrier has typically much larger scale. Indeed, its spatial scale is about $r_b \sim m\gamma^2/\omega^2$ where m is effective mass of the exciton, γ is the exciton-phonon coupling constant, and ω is the characteristic frequency of optical phonons. Excitons are self-trapped when m and γ are large, and then the spatial size of the barrier is large compared with the lattice spacing. Transforming a free exciton state into a self-trapped one proceeds as a collective tunneling of coupled exciton-lattice system (an instanton). Because r is large, tunneling can be described by a continuum theory. The height of the barrier $W \sim \omega^4/m^3\gamma^4$. Because both m and γ appear in the denominator of w, the barriers are basically low. Therefore, free excitons can be seen in crystals with strong exciton-phonon coupling only in pure samples and at low temperatures. Coexistence of free and self-trapped excitons was observed in rare-gas solids, alkali-halides, and in molecular crystal of pyrene.

Interaction

Excitons are the main mechanism for light emission in semiconductors at low temperature (when the characteristic thermal energy kT is less than the exciton binding energy), replacing the free electron-hole recombination at higher temperatures.

The existence of exciton states may be inferred from the absorption of light associated with their excitation. Typically, excitons are observed just below the band gap.

When excitons interact with photons a so-called polariton (also exciton-polaritons) is formed. These excitons are sometimes referred to as *dressed excitons*.

Provided the interaction is attractive, an exciton can bind with other excitons to form a biexciton, analogous to a dihydrogen molecule. If a large density of excitons is created in a material, they can interact with one another to form an electron-hole liquid, a state observed in k-space indirect semiconductors.

Additionally, excitons are integer-spin particles obeying Bose statistics in the low-density limit. In some systems, where the interactions are repulsive, a Bose–Einstein condensed state is predicted to be the ground state. Exciton condensates have been seen in a double quantum well systems.

Spatially Direct and Indirect Excitons

Normally, excitons in a semiconductor have a very short lifetime due to the close proximity of the electron and hole. However, by placing the electron and hole in spatially separated quantum wells with an insulating barrier layer in between so called 'spatially indirect' excitons can be created. In contrast to ordinary (spatially direct), these spatially indirect excitons can have large spatial separation between the electron and hole, and thus possess a much longer lifetime. This is often used to cool excitons to very low temperatures in order to study Bose Einstein condensation (or rather its 2 dimensional analog).

MAGNON

A magnon is a quasiparticle, a collective excitation of the electrons' spin structure in a crystal lattice. In the equivalent wave picture of quantum mechanics, a magnon can be viewed as a quantized spin wave. Magnons carry a fixed amount of energy and lattice momentum, and are spin-1, indicating they obey boson behavior.

Brief History

The concept of a magnon was introduced in 1930 by Felix Bloch in order to explain the reduction of the spontaneous magnetization in a ferromagnet. At absolute zero temperature, a Heisenberg ferromagnet reaches the state of lowest energy, in which all of the atomic spins (and hence magnetic moments) point in the same direction. As the temperature increases, more and more spins deviate randomly from the alignment, increasing the internal energy and reducing the net magnetization. If one views the perfectly magnetized state at zero temperature as the vacuum state of the ferromagnet, the low-temperature state with a few misaligned spins can be viewed as a gas of quasiparticles, in this case magnons. Each magnon reduces the total spin along the direction of magnetization by one unit of $\hbar$ and the magnetization by $\gamma\hbar$, where γ is the gyromagnetic ratio. This leads to Bloch›s law for the temperature dependence of spontaneous magnetization:

$$M(T) = M_0(1 - (T/T_C)^{3/2})$$

The quantitative theory of magnons, quantized spin waves, was developed further by Theodore Holstein, Henry Primakoff (1940), and Freeman Dyson (1956). Using the second quantization formalism they showed that magnons behave as weakly interacting quasiparticles obeying Bose–Einstein statistics (bosons). A comprehensive treatment can be found in the solid state textbook by Charles Kittel or the early review article by Van Kranendonk and Van Vleck.

Direct experimental detection of magnons by inelastic neutron scattering in ferrite was achieved in 1957 by Bertram Brockhouse. Since then magnons have been detected in ferromagnets, ferrimagnets, and antiferromagnets.

The fact that magnons obey the Bose–Einstein statistics was confirmed by the light scattering experiments done during the 1960s through the 1980s. Classical theory predicts equal intensity of Stokes and anti-Stokes lines. However, the scattering showed that if the magnon energy is comparable to or smaller than the thermal energy, or $\hbar\omega < k_BT$, then the Stokes line becomes more intense, as follows from Bose–Einstein statistics. Bose–Einstein condensation of magnons was proven in an antiferromagnet at low temperatures by Nikuni *et al.* and in a ferrimagnet by Demokritov *et al.* at room temperature. Recently Uchida *et al.* reported the generation of spin currents by surface plasmon resonance.

Properties

Magnon behavior can be studied with a variety of scattering techniques. Magnons behave as a Bose gas with no chemical potential. Microwave pumping can be used to excite spin waves and create additional non-equilibrium magnons which thermalize into phonons. At a critical density, a condensate is formed, which appears as the emission of monochromatic microwaves. This microwave source can be tuned with an applied magnetic field.

PHONON

In physics, a plasmon is a quantum of plasma oscillation. Just as light [optical oscillation] consists of photons, the plasma oscillation consists of plasmons. The plasmon can be considered as a quasiparticle since it arises from the quantization of plasma oscillations, just like phonons are quantizations of mechanical vibrations. Thus, plasmons are collective (a discrete number) oscillations of the free electron gas density. For example, at optical frequencies, plasmons can couple with a photon to create another quasiparticle called a plasmon polariton.

Derivation

The plasmon was initially proposed in 1952 by David Pines and David Bohm and was shown to arise from a Hamiltonian for the long-range electron-electron correlations.

Since plasmons are the quantization of classical plasma oscillations, most of their properties can be derived directly from Maxwell's equations.

Explanation

Plasmons can be described in the classical picture as an oscillation of free electron density with respect to the fixed positive ions in a metal. To visualize a plasma oscillation, imagine a cube of metal placed in an external electric field pointing to the right. Electrons will move to the left side (uncovering positive ions on the right side) until they cancel the field inside the metal. If the electric field is removed, the electrons move to the right, repelled by each other and attracted to the positive ions left bare on the right side. They oscillate back and forth at the plasma frequency until the energy is lost in some kind of resistance or damping. Plasmons are a quantization of this kind of oscillation.

Role of Plasmons

Plasmons play a large role in the optical properties of metals and semiconductors. Light of frequencies below the plasma frequency is reflected by a material because the electrons in the material screen the electric field of the light. Light of frequencies above the plasma frequency is transmitted by a material because the electrons in the material cannot respond fast enough to screen it. In most metals, the plasma frequency is in the ultraviolet, making them shiny (reflective) in the visible range. Some metals, such as copper and gold, have electronic interband transitions in the visible range, whereby specific light energies (colors) are absorbed, yielding their distinct color. In semiconductors, the valence electron plasmon frequency is usually in the deep ultraviolet, while their electronic interband transitions are in the visible range, whereby specific light energies (colors) are absorbed, yielding their distinct color which is why they are reflective. It has been shown that the plasmon frequency may occur in the mid-infrared and near-infrared region when semiconductors are in the form of nanoparticles with heavy doping.

The plasmon energy can often be estimated in the free electron model as

$$E_p = \hbar\sqrt{\frac{ne^2}{m\epsilon_0}} = \hbar\omega_p,$$

where n is the conduction electron density, e is the elementary charge, m is the electron mass, ϵ_0 the permittivity of free space, $\hbar$ the reduced Planck constant and ω_p the plasmon frequency.

Surface Plasmons

Surface plasmons are those plasmons that are confined to surfaces and that interact strongly with light resulting in a polariton. They occur at the interface of a material exhibiting positive real part of their relative permittivity, i.e. dielectric constant, (e.g. vacuum, air, glass and other dielectrics) and a material whose real part of permittivity is negative at the given frequency of light, typically a metal or heavily doped semiconductors. In addition to opposite sign of the real part of the permittivity, the magnitude of the real part of the permittivity in the negative permittivity region should typically be larger than the magnitude of the permittivity in the positive permittivity region, otherwise the light is not bound to the surface (i.e. the surface plasmons do not exist) as shown in the famous book by Raether. At visible wavelengths of light, e.g. 632.8 nm wavelength provided by a He-Ne laser, interfaces supporting surface plasmons are often formed by metals like silver or gold (negative real part permittivity) in contact with dielectrics such as air or silicon dioxide. The particular choice of materials can have a drastic effect on the degree of light confinement and propagation distance due to losses. Surface plasmons can also exist on interfaces other than flat surfaces, such as particles, or rectangular strips, v-grooves, cylinders, and other structures. Many structures have been investigated due to the capability of surface plasmons to confine light below the diffraction limit of light.

Surface plasmons can play a role in surface-enhanced Raman spectroscopy and in explaining anomalies in diffraction from metal gratings (Wood's anomaly), among other things. Surface plasmon resonance is used by biochemists to study the mechanisms and kinetics of ligands binding to receptors (i.e. a substrate binding to an enzyme). Multi-Parametric Surface Plasmon Resonance can be used not only to measure molecular interactions, but also nanolayer properties or structural changes in the adsorbed molecules, polymer layers or graphene, for instance.

Surface plasmons may also be observed in the X-ray emission spectra of metals. A dispersion relation for surface plasmons in the X-ray emission spectra of metals has been derived (Harsh and Agarwal).

More recently surface plasmons have been used to control colors of materials. This is possible since controlling the particle's shape and size determines the types of surface plasmons that can couple to it and propagate across it. This in turn controls the interaction of light with the surface. These effects are illustrated by the historic stained glass which adorn medieval cathedrals. In this case, the color is given by metal nanoparticles of a fixed size which interact with the optical field to give the glass its vibrant color. In modern science, these effects have been engineered for both visible light and microwave radiation. Much research goes on first in the microwave range because at this wavelength material surfaces can be produced mechanically as the patterns tend to be of the order a few centimeters. To produce optical range surface plasmon effects involves producing surfaces which have features <400 nm. This is much more difficult and has only recently become possible to do in any reliable or available way.

Gothic stained glass rose window of Notre-Dame de Paris. The colors were achieved by colloids of gold nano-particles.

Recently, graphene has also been shown to accommodate surface plasmons, observed via near field infrared optical microscopy techniques and infrared spectroscopy. Potential applications of graphene plasmonics mainly addressed the terahertz to midinfrared frequencies, such as optical modulators, photodetectors, biosensors.

Possible Applications

The position and intensity of plasmon absorption and emission peaks are affected by molecular adsorption, which can be used in molecular sensors. For example, a fully operational device detecting casein in milk has been prototyped, based on detecting a change in absorption of a gold layer. Localized surface plasmons of metal nanoparticles can be used for sensing different types molecules, proteins, etc.

Plasmons are being considered as a means of transmitting information on computer chips, since plasmons can support much higher frequencies (into the 100 THz range, whereas conventional wires become very lossy in the tens of GHz). However, for plasmon-based electronics to be practical, a plasmon-based amplifier analogous to the transistor, called a plasmonstor, needs to be created.

Plasmons have also been proposed as a means of high-resolution lithography and microscopy due to their extremely small wavelengths; both of these applications have seen successful demonstrations in the lab environment.

Finally, surface plasmons have the unique capacity to confine light to very small dimensions, which could enable many new applications.

Surface plasmons are very sensitive to the properties of the materials on which they propagate. This has led to their use to measure the thickness of monolayers on colloid films, such as screening and quantifying protein binding events. Companies such as Biacore have commercialized instruments that operate on these principles. Optical surface plasmons are being investigated with a view to improve makeup by L'Oréal and others.

In 2009, a Korean research team found a way to greatly improve organic light-emitting diode efficiency with the use of plasmons.

A group of European researchers led by IMEC has begun work to improve solar cell efficiencies and costs through incorporation of metallic nanostructures (using plasmonic effects) that can enhance absorption of light into different types of solar cells: crystalline silicon (c-Si), high-performance III-V, organic, and dye-sensitized. However, for plasmonic photovoltaic devices to function optimally, ultra-thin transparent conducting oxides are necessary. Full color holograms using *plasmonics* have been demonstrated.

"The silicon absorbs only a certain fraction of the spectrum, and it's transparent to the rest. If I put a photo voltaic module on my roof, the silicon absorbs that portion of the spectrum, and some of that light gets converted into power, but the rest of it ends up just heating the roof and waste. The new technique they've developed is based on a phenomenon observed in metallic structures known as plasmon resonance. Plasmons are coordinated waves, or ripples, of electrons that exist on the surfaces of metals at the point where the metal meets the air.

POLARITON

In physics, polaritons are quasiparticles resulting from strong coupling of electromagnetic waves with an electric or magnetic dipole-carrying excitation. They are an expression of the common quantum phenomenon known as level repulsion, also known as the avoided crossing principle.

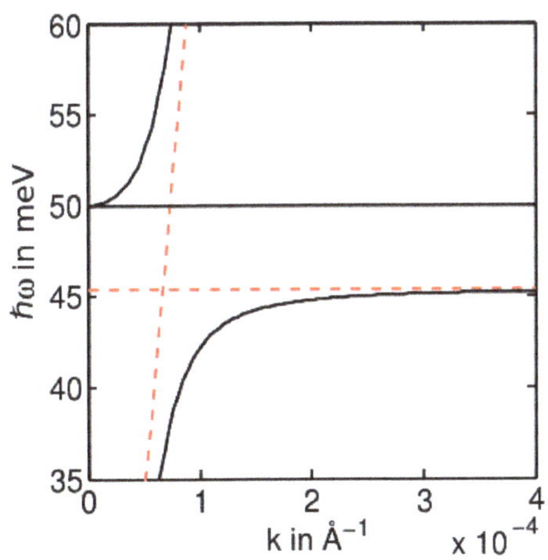

Dispersion relation of polaritons in GaP. Red curves are the uncoupled phonon and photon dispersion relations, black curves are the result of coupling (from top to bottom: upper polariton, LO phonon, lower polariton).

Polaritons describe the crossing of the dispersion of light with any interacting resonance. The polariton is a bosonic quasiparticle, and should not be confused with the polaron (a fermionic one), which is an electron plus an attached phonon cloud.

Whenever the polariton picture is valid, the model of photons propagating freely in crystals is insufficient. A major feature of polaritons is a strong dependency of the propagation speed of light through the crystal on the frequency. For exciton-polaritons, rich experimental results on various aspects have been gained in copper (I) oxide.

History

Oscillations in ionized gases were observed by Tonks and Langmuir in 1929. Polaritons were first considered theoretically by Tolpygo. They were termed light-excitons in Ukrainian and Russian scientific literature. That name was suggested by Pekar, but the term *polariton,* proposed by Hopfield, was adopted. Coupled states of electromagnetic waves and phonons in ionic crystals and their dispersion relation, now known as phonon polaritons, were obtained by Tolpygo in 1950and, independently, by Kun in 1951. Collective interactions were published by Pines and Bohm in 1952, and plasmons were described in silver by Fröhlich and Pelzer in 1955. Ritchie predicted surface plasmons in 1957, then Ritchie and Eldridge published experiments and predictions of emitted photons from irradiated metal foils in 1962. Otto first published on surface plasmon-polaritons in 1968.

Types

A polariton is the result of the mixing of a photon with a polar excitation of a material. The following are types of polaritons:

- Phonon polaritons result from coupling of an infrared photon with an optic phonon;
- Exciton polaritons result from coupling of visible light with an exciton;
- Intersubband polaritons result from coupling of an infrared or terahertz photon with an intersubband excitation;
- Surface plasmon polaritons result from coupling of surface plasmons with light (the wavelength depends on the substance and its geometry);
- Bragg polaritons (“Braggoritons”) result from coupling of Bragg photon modes with bulk excitons.
- Plexcitons result from coupling plasmons with excitons.

TRION

A trion is a localized excitation which consists of three charged quasiparticles. A negative trion consists of two electrons and one hole and a positive trion consists of two holes and one electron. The trion itself is a quasiparticle and is somewhat similar to an exciton, which is a complex of one electron and one hole. The trion has a ground singlet state (spin $s = 1/2$) and an excited triplet state ($s = 3/2$). Here singlet and triplet degeneracies originate not from the whole system but from the two identical particles in it. The half-integer spin value distinguishes trions from excitons in many phenomena; for example, energy states of trions, but not excitons, are split in an applied magnetic field. Trion states were predicted theoretically and then observed experimentally in various optically excited semiconductors, especially in quantum dots and quantum well structures. There is evidence of their existence in nanotubes. Despite numerous reports of experimental trion observations in different semiconductor heterostructures, there are serious concerns on the exact physical nature of the detected complexes. The originally foreseen 'true' trion particle has a delocalized wavefunction (at least at the scales of several Bohr radii) while recent studies reveal significant binding from charged impurities in real semiconductor quantum wells

Trions have been observed in atomically thin two-dimensional (2D) transition-metal dichalcogenide semiconductors. In 2D materials the form of the interaction between charge carriers is modified by the nonlocal screening provided by the atoms in the layer. The interaction is approximately logarithmic at short range and of Coulomb $1/r$ form at long range. The diffusion Monte Carlo method has been used to obtain numerically exact results for the binding energies of trions in 2D semiconductors within the effective mass approximation,

REFERENCES

1. E. Kaxiras, Atomic and Electronic Structure of Solids, ISBN 0-521-52339-7, pages 65–69.
2. A guide to Feynman diagrams in the many-body problem, by Richard D. Mattuck, p10.
3. Principles of Nanophotonics by Motoichi Ohtsu, p205 google books link
4. A. Gelfert, 'Manipulative Success and the Unreal', International Studies in the Philosophy of Science Vol. 17, 2003, 245–263
5. B.Falkenburg, Particle Metaphysics (The Frontiers Collection), Berlin: Springer 2007, esp. pp. 243–46
6. J. E. Hoffman; McElroy, K; Lee, DH; Lang, KM; Eisaki, H; Uchida, S; Davis, JC; et al. (2002). "Imaging Quasiparticle Interference in Bi2Sr2CaCu2O8+". Science. 297 (5584): 1148–51. arXiv:cond-mat/0209276Freely accessible. Bibcode:2002Sci...297.1148H. doi:10.1126/science.1072640. PMID 12142440.
7. Banerjee, A.; Bridges, C. A.; Yan, J.-Q.; et al. (4 April 2016). "Proximate Kitaev quantum spin liquid behaviour in a honeycomb magnet". Nature Materials. doi:10.1038/nmat4604. (subscription required (help)).
8. Shaginyan, V. R.; et al. (2012). "Identification of Strongly Correlated Spin Liquid in Herbertsmithite". EPL. 97 (5): 56001. arXiv:1111.0179Freely accessible. Bibcode:2012EL.....9756001S. doi:10.1209/0295-5075/97/56001.
9. R. S. Knox, Theory of excitons, Solid state physics (Ed. by Seitz and Turnbul, Academic, NY), v. 5, 1963.
10. Liang, W Y (1970). "Excitons". Physics Education. 5 (125301).

Bibcode:1970PhyEd...5..226L. doi:10.1088/0031-9120/5/4/003.

11. Couto, ODD; Puebla, J (2011). "Charge control in InP/(Ga,In)P single quantum dots embedded in Schottky diodes". Physical Review B. 84 (4): 226. Bibcode:1970PhyEd...5..226L. doi:10.1103/PhysRevB.84.125301.
12. Frenkel, J. (1931). "On the Transformation of light into Heat in Solids. I". Physical Review. 37: 17. Bibcode:1931PhRv...37...17F. doi:10.1103/PhysRev.37.17.
13. Wannier, Gregory (1937). "The Structure of Electronic Excitation Levels in Insulating Crystals". Physical Review. 52 (3): 191. Bibcode:1937PhRv...52..191W. doi:10.1103/PhysRev.52.191.
14. J. D. Wright (1995) Molecular Crystals (2nd ed.). Cambridge University Press. p. 108. ISBN 0-521-47730-1.
15. Ivan Pelant, Jan Valenta (2012). Luminescence Spectroscopy of Superconductors. Oxford University Press. p. 161. ISBN 978-0-19-958833-6.
16. Guglielmo Lanzani (2012). The Photophysics Behind Photovoltaics and Photonics. Wiley-VCH Verlag. p. 82.
17. A. Prikhotjko, Absorption Spectra of Crystals at Low Temperatures, J. Physics USSR 8, 257 (1944)
18. F. Prikhot'ko, Izv, AN SSSR Ser. Fiz. 7, 499 (1948) http://ujp.bitp.kiev.ua/files/journals/53/si/53SI18p.pdf
19. V. L. Broude, E. I. Rashba, and E. F. Sheka, Spectroscopy of molecular excitons (Springer, NY) 1985
20. E. I. Rashba, Giant Oscillator Strengths Associated with Exciton Complexes, Sov. Phys. Semicond. 8, 807-816 (1975)
21. N. Schwentner, E.-E. Koch, and J. Jortner, Electronic excitations in condensed rare gases, Springer tracts in modern physics, 107, 1 (1985).
22. M. Ueta, H. Kanzaki, K. Kobayashi, Y. Toyozawa, and E. Hanamura. Excitonic Processes in Solids, Springer Series in Solid State Sciences, Vol. 60 (1986).
23. E. I. Rashba, "Theory of Strong Interaction of Electron Excitations with Lattice Vibrations in Molecular Crystals, Optika i Spektroskopiya 2, 75, 88 (1957).
24. E. I. Rashba, Self-trapping of excitons, in: Excitons (North-Holland, Amsterdam, 1982), p. 547.
25. A. S. Ioselevich and E. I. Rashba, Theory of Nonradiative Trapping in Crystals, in: "Quantum tunneling in condensed media." Eds. Yu. Kagan and A. J. Leggett. (North-Holland, Amsterdam, 1992), p. 347-425.
26. G. Zimmerer, "Excited-State Spectroscopy in Solids", in: Proceedings of the International School of Physics,``Enrico Fermi, Course XCVI, Varenna, Italy, 1985." (1987).
27. Ya. Fugol', "Free and self-trapped excitons in cryocrystals: kinetics and relaxation processes." Advances in Physics 37, 1-35 (1988).
28. Ch. B. Lushchik, in "Excitons," edited by E. I. Rashba, and M. D. Sturge, (North Holland, Amsterdam, 1982), p. 505.

29. M. Furukawa, Ken-ichi Mizuno, A. Matsui, N. Tamai and I. Yamazaiu, Branching of Exciton Relaxation to the Free and Self-Trapped Exciton States, Chemical Physics 138, 423 (1989).

30. "Exciton Condensation in Bilayer Quantum Hall Systems". Annual Review of Condensed Matter Physics. January 10, 2014. doi:10.1146/annurev-conmatphys-031113-133832.

31. A.A. High (2012) "Spontaneous coherence in a cold exciton gas" Nature

Index